Nick Vandome

macOS Monterey

for all Macs (iMac, Mac mini, Mac Pro and MacBook)
with macOS Monterey

In easy steps is an imprint of In Easy Steps Limited
16 Hamilton Terrace · Holly Walk · Leamington Spa
Warwickshire · United Kingdom · CV32 4LY
www.ineasysteps.com

In Easy Steps Limited supports The Forest Stewardship Council (FSC), the leading international forest certification organization. All our titles that are printed on Greenpeace approved FSC certified paper carry the FSC logo.

MIX
Paper from responsible sources
FSC® C020837

Printed and bound in the United Kingdom

ISBN 978-1-84078-946-1

Contents

1 Introducing macOS Monterey

macOS Monterey is the latest operating system from Apple, for its range of desktop and laptop computers. This chapter introduces macOS Monterey, showing how to install it and become familiar with the user interface and some of the features needed to get up and running.

About macOS Monterey

macOS Monterey is the latest version of the operating system for Apple computers: the iMac, MacBook, Mac mini and Mac Pro. It is the first version of the operating system to have the 12 designation, after over a decade of versions with the 10 designation and then macOS Big Sur 11. The change from version 10 was primarily to recognize the fact that Mac desktops and laptops started using their own M1 processing chip, in a move away from Intel ones. macOS is still based on the UNIX programming language, which is a very stable and secure operating environment, and this ensures that macOS is secure, user-friendly and stylish.

macOS Monterey continues the functionality of several of its Mac operating system predecessors, including using some of the functionality that is available on Apple's mobile devices such as the iPad and the iPhone. The two main areas where the functionality of the mobile devices has been incorporated are:

- The way apps can be downloaded and installed. Instead of using a disc, macOS Monterey utilizes the App Store to provide apps, which can be installed in a couple of steps.
- Using Multi-Touch Gestures, with a laptop trackpad, Apple's Magic Trackpad, or Magic Mouse, for navigating macOS.

Some of the new and updated features of macOS Monterey are:

- **FaceTime**. A major upgrade for the FaceTime video-chatting app, which now includes options for sharing movies and music with other people on a FaceTime call, and also sharing your screen on a call.
- **Quick Notes**. The Notes app now includes a Quick Notes option, where notes can be created from any screen, and then saved and accessed in the Notes app.
- **Focus**. The Focus option can be used to set conditions to limit the number of notifications you receive when you are performing specific tasks.
- **iCloud+**. The online storage facility, iCloud, has been updated with iCloud+, to give even more options.
- **Enhanced apps**. macOS Monterey has enhanced versions of Safari, Messages and Maps.

UNIX is an operating system that has traditionally been used for large commercial mainframe computers. It is renowned for its stability and ability to be used within different computing environments.

The Apple Pencil can be used on some Mac computers with macOS Monterey to perform tasks such as drawing and sketching, in compatible apps, and marking up screenshots and PDF documents.

The New icon pictured above indicates a new or enhanced feature introduced with the latest version of macOS – macOS Monterey.

Installing macOS Monterey

When it comes to installing macOS Monterey you do not need to worry about an installation CD or DVD; it can be downloaded and installed directly from the online App Store. New Macs will have macOS Monterey installed. The following range is compatible with macOS Monterey and can be upgraded with it, directly from the App Store:

- iMac (Late 2015 or newer).
- iMac Pro (2017 or newer).
- MacBook (Early 2016 or newer).
- MacBook Pro (Early 2015 or newer).
- MacBook Air (Early 2015 or newer).
- Mac mini (Late 2014 or newer).
- Mac Pro (Late 2013 or newer).

If you want to install macOS Monterey on an existing Mac, you will need to have the minimum requirements of:

- 2GB of memory.
- 12.5GB of available storage for installation.

If your Mac meets these requirements, you can download and install macOS Monterey, for free, as follows:

1. Click on this icon on the Dock to access the App Store (or select **Software Update...**; see the second Hot tip)

2. Locate the **macOS Monterey** icon using the App Store Search box

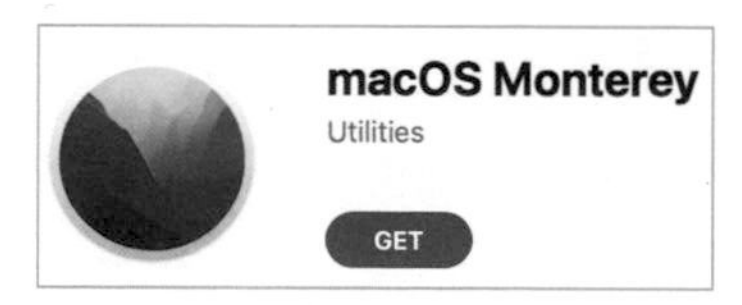

3. Click on the **Get** button and follow the installation instructions

macOS Monterey is a new version of the Mac operating system. It contains a number of enhancements to make the operating system faster and more reliable than ever.

For details about using the Dock, see pages 26-33.

To check your computer's software version and upgrade options, click on **Apple menu** > **About This Mac** from the main Menu bar. Click on the **Overview** tab and click on the **Software Update...** button. See page 13 for details. Also, download any updates from **System Preferences** > **Software Update**.

The macOS Environment

The first most noticeable element about macOS is its elegant user interface. This has been designed to create a user-friendly graphic overlay to the UNIX operating system at the heart of macOS, and it is a combination of rich colors and sharp, original graphics. The main elements that make up the initial macOS environment are:

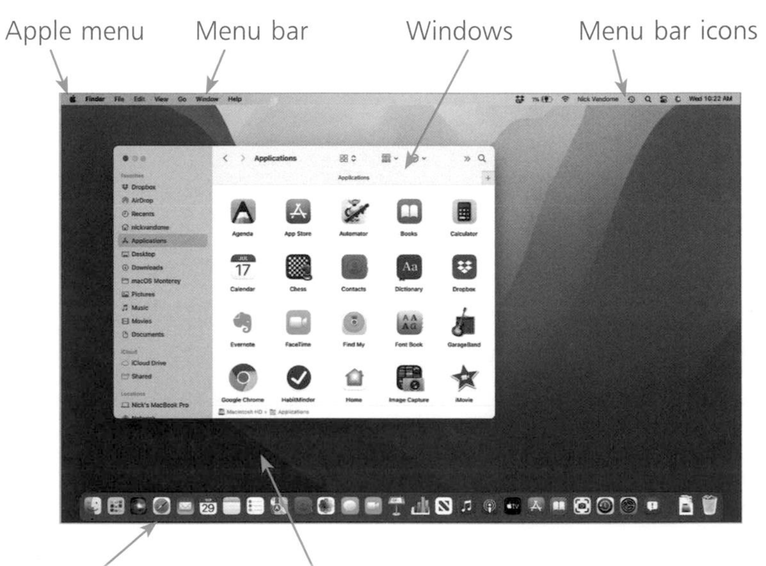

The Dock is designed to help make organizing and opening items as quick and easy as possible. For a detailed look at the Dock, see Chapter 2.

Many of the behind-the-scenes features of macOS Monterey are aimed at saving power on your Mac. These include timer-coalescing technologies for saving processing and battery power; features for saving energy when apps are not being used; power-saving features in Safari for ignoring additional content provided by web page plug-ins; and memory compression to make your Mac quicker and more responsive.

The **Apple menu** is standardized throughout macOS, regardless of the app in use.

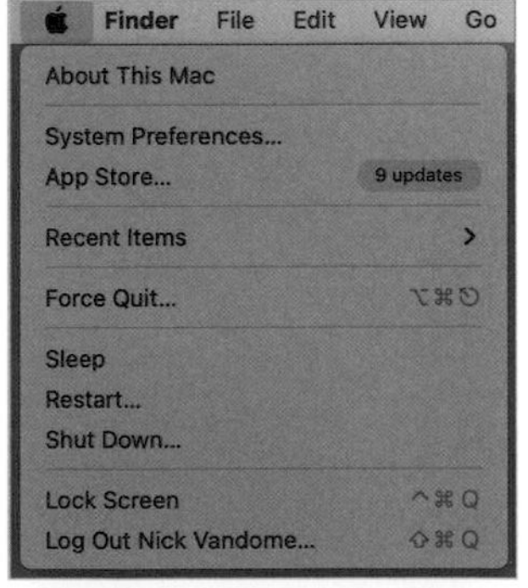

Menus

Menus in macOS contain commands for the operating system and any relevant apps. If there is an arrow next to a command it means there are subsequent options for the item. Some menus also incorporate the same transparency as the sidebar so that the background shows through.

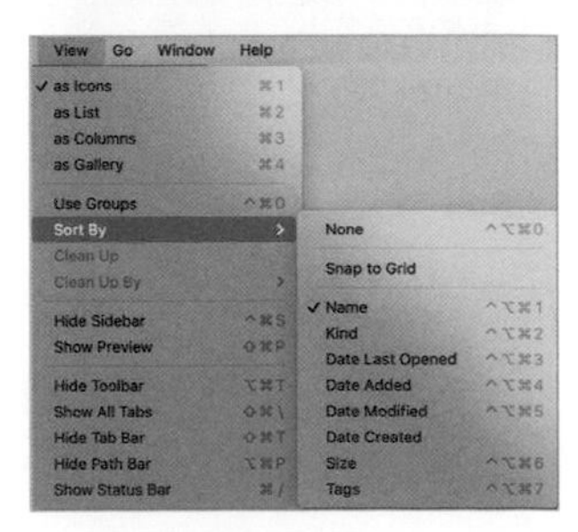

Transparency

One feature in macOS Monterey is that the sidebar and toolbars in certain apps are transparent so that you can see some of the Desktop behind it. This also helps the uppermost window blend in with the background.

1. In certain apps with a sidebar, such as the Finder or Safari, the background appears behind the sidebar
2. When you move the window, the background behind the sidebar changes accordingly

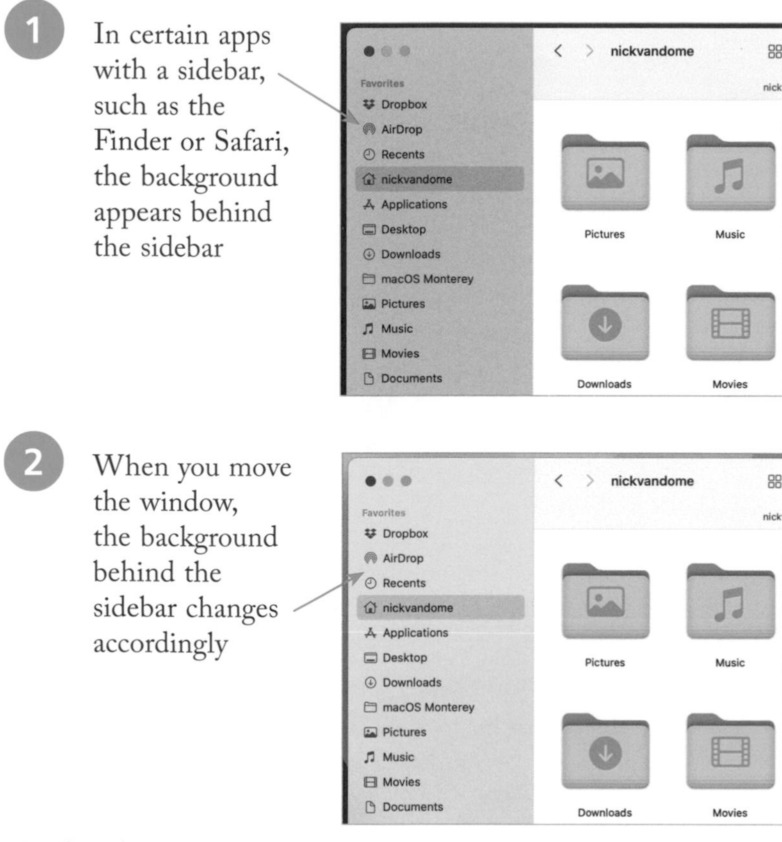

Window buttons

These appear in any open macOS window and can be used to manipulate the window. They include a full-screen option. Use the window buttons to, from left to right: close a window, minimize a window, or access full-screen mode (if available).

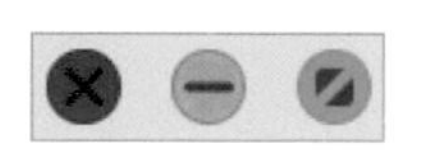

If an app has full-screen functionality, click on this button to return to standard view:

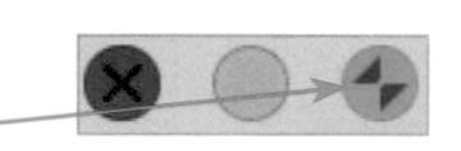

If an option is gray, it means it is not available; i.e. the screen cannot be maximized.

The red window button is used to close a window, and in some cases – such as with Notes, Reminders and Calendar – it also closes the app. The amber button is used to minimize a window so that it appears at the right-hand side of the Dock.

About Your Mac

When you buy a new Mac, you will almost certainly check the technical specifications before you make a purchase. Once you have your Mac, there will be times when you will want to view these specifications again, such as the version of macOS in use, the amount of memory and the amount of storage. This can be done through the About This Mac option, which can be accessed from the Apple menu. To do this:

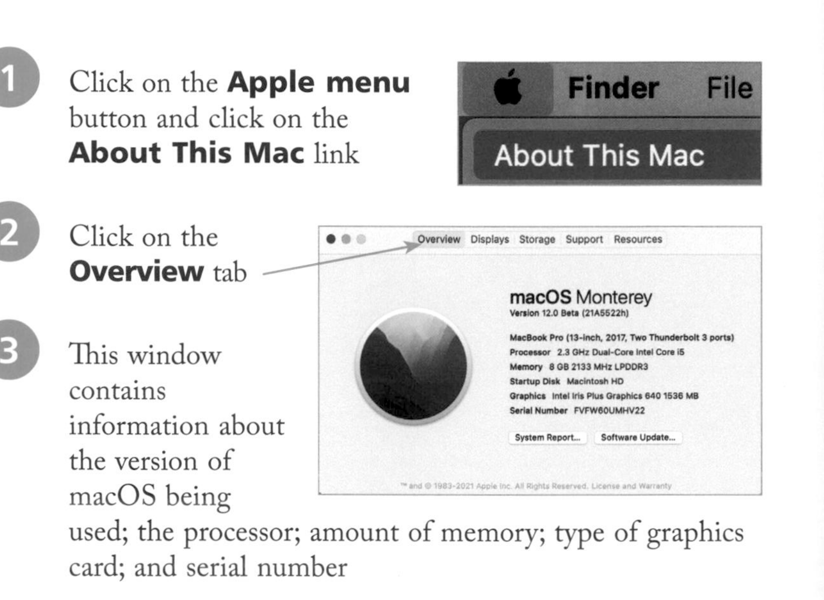

1. Click on the **Apple menu** button and click on the **About This Mac** link

2. Click on the **Overview** tab

3. This window contains information about the version of macOS being used; the processor; amount of memory; type of graphics card; and serial number

4. Click on the **System Report...** button to view full details about the hardware and software on your Mac

Click on the **Software Update...** button to see available software updates for your Mac

Display information

This gives information about your Mac's display.

Click on the **Displays** tab

2. This window contains information about your Mac's display, including the type, size, resolution, and graphics card

3. Click on the **Displays Preferences...** button to view options for changing the display's resolution, brightness and color

For more information about software updates, see page 181.

...cont'd

Storage information

This contains information about your Mac's physical and removable storage.

1. Click on the **Storage** tab Storage

2. This window contains information about the used and available storage on your hard disk, and also options for writing various types of CDs and DVDs (if applicable)

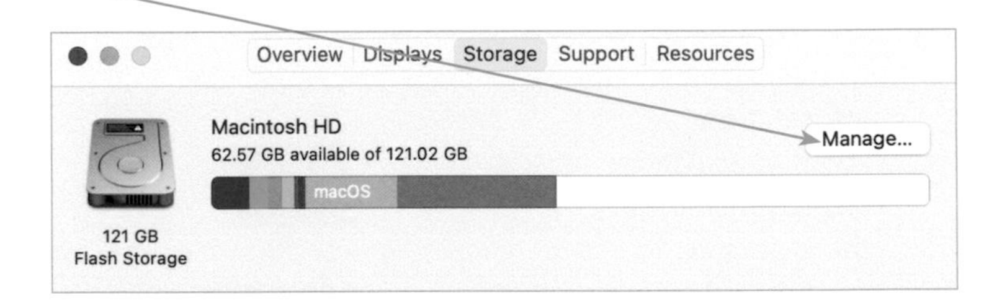

Click on the **Manage...** button in the Storage window to view how much space specific apps are taking up, and also options for optimizing storage on your Mac. With macOS Monterey, this is known as **Optimized Storage**. Optimized Storage includes a **Store in iCloud** option, which can be used to enable your Mac to identify files that have not been opened or used in a long time, and then automatically store them in iCloud (if you have enough storage space there). Their location remains the same on your Mac, but they are physically kept in iCloud, thus freeing up more space on your Mac. Optimized Storage also has options for removing movies and TV shows that have been watched, emptying the Trash automatically and reducing overall clutter on your Mac.

Managing memory

This contains information about your Mac's memory, which is used to run macOS and also the applications on your computer.

1. Click on the **Manage...** button Manage...

2. This window contains information about how memory can be managed in macOS Monterey (see the Don't forget tip)

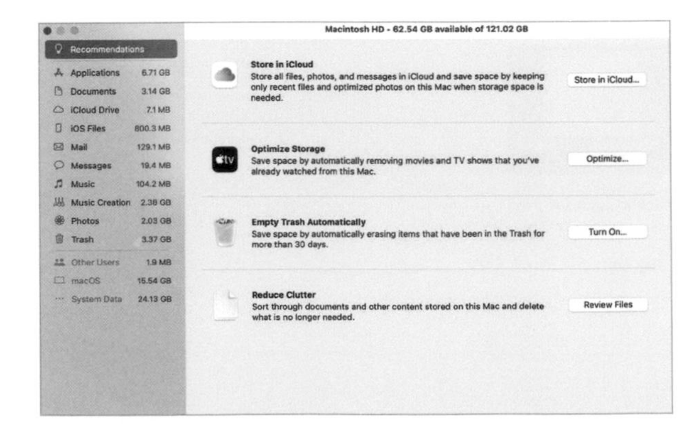

Support

The **Support** tab provides links to a range of help options for your Mac and macOS.

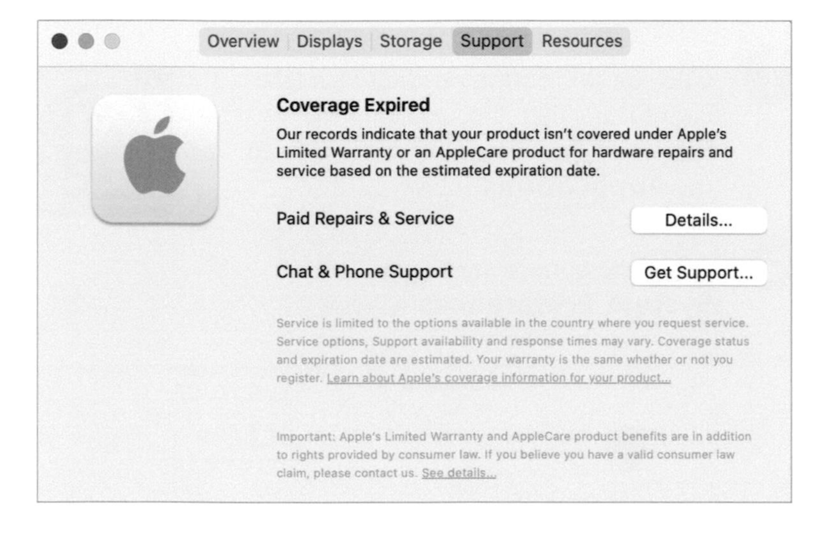

Resources

The **Resources** tab provides links to a range of help options for your Mac and macOS.

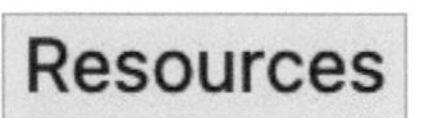

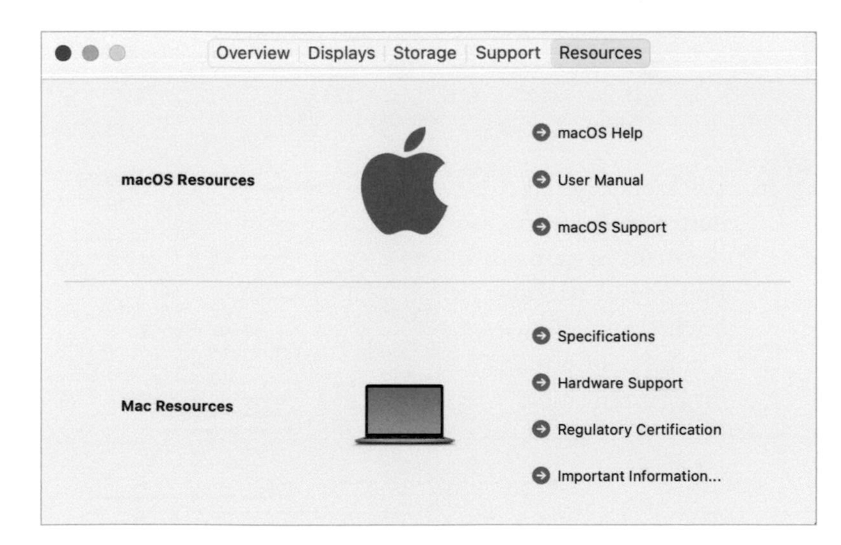

The **Resources** section provides links to online resources for a variety of areas related to Mac computers and macOS.

About System Preferences

macOS Monterey has a wide range of options for customizing and configuring the way that your Mac operates. These are located within the System Preferences section. To access this:

For more detailed information about the Dock, see Chapter 2.

For a detailed look at System Preferences, see pages 34-35.

The **General** option in System Preferences can be used to change the overall appearance of macOS Monterey, including colors for buttons, menus, scroll bars and windows. Another option in System Preferences is for **Desktop & Screen Saver**. This has tabs for these items, from which you can select a background wallpaper for your Mac, including the Dynamic Desktop, and also a screen saver for when it is inactive.

1. Click on this button on the Dock (the bar of icons that appears along the bottom of the screen), or access it from the **Applications** folder

2. All of the options are shown in the **System Preferences** window

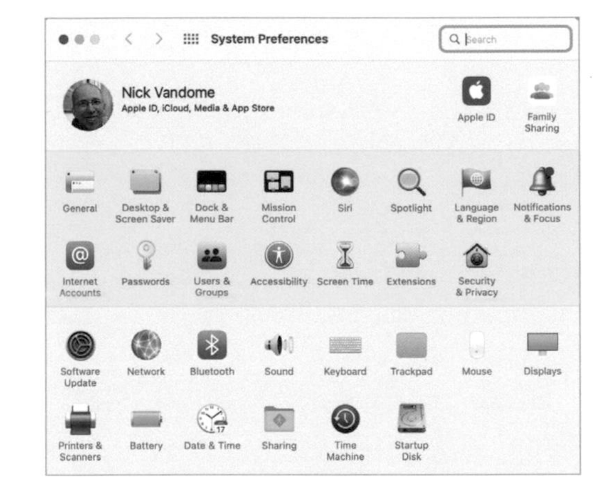

3. Click once on an item to open it in the main System Preferences window. Each item will have a number of options for customization

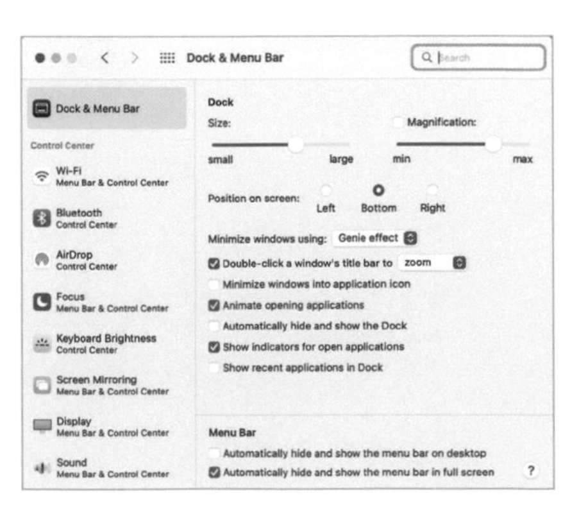

4. Click on the **Show All** button to return to the main System Preferences window

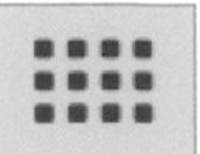

Search Options

Finding items on a Mac or on the web is an important part of everyday computing, and with macOS Monterey there are a number of options for doing this.

Siri

Siri is Apple's digital voice assistant that can search for items using speech. It has been available on Apple's mobile devices using iOS for a number of years, and with macOS Monterey it is available on Macs and MacBooks. See pages 19-21 for more details about setting up and using Siri.

Spotlight search

Spotlight is the dedicated search app for macOS. It can be used over all the files on your Mac and the internet. To use Spotlight:

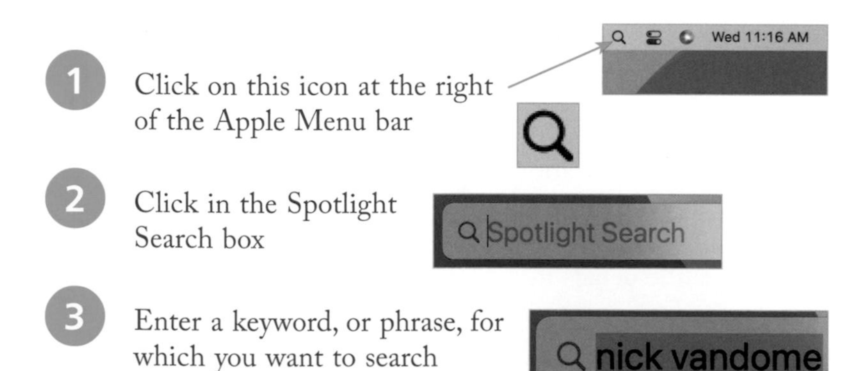

1. Click on this icon at the right of the Apple Menu bar
2. Click in the Spotlight Search box
3. Enter a keyword, or phrase, for which you want to search
4. The top hits from within your apps, and the web, are shown in the left-hand panel. Click on an item to view its details

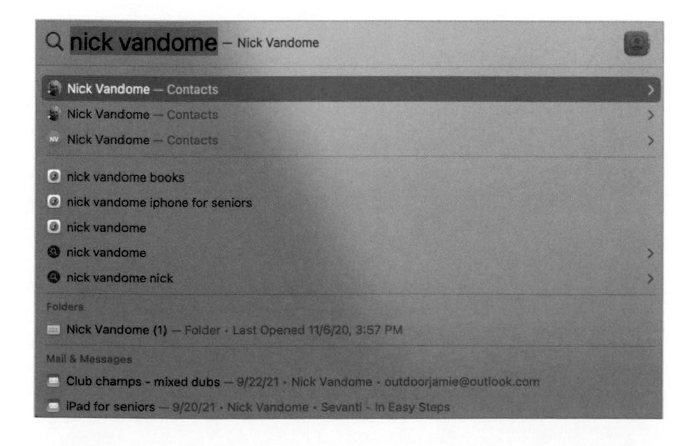

Spotlight starts searching for items as soon as you start typing a word. So don't worry if some of the first results look inappropriate, as these will disappear once you have finished typing the full word.

...cont'd

For items linked to apps on your Mac, click on an item to view the information in the relevant app

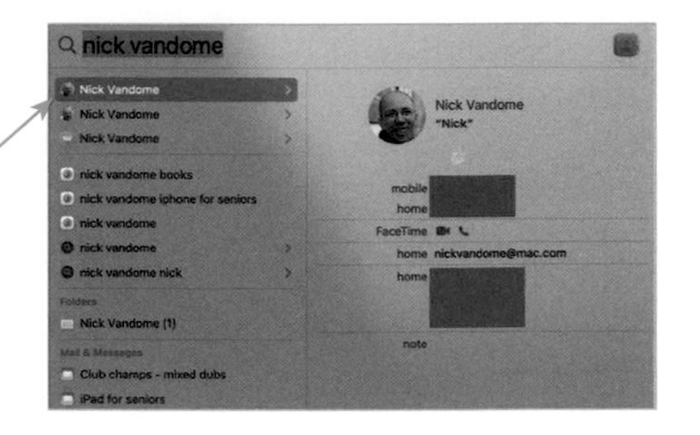

6 For items from the web, click on an item to view the related web page(s)

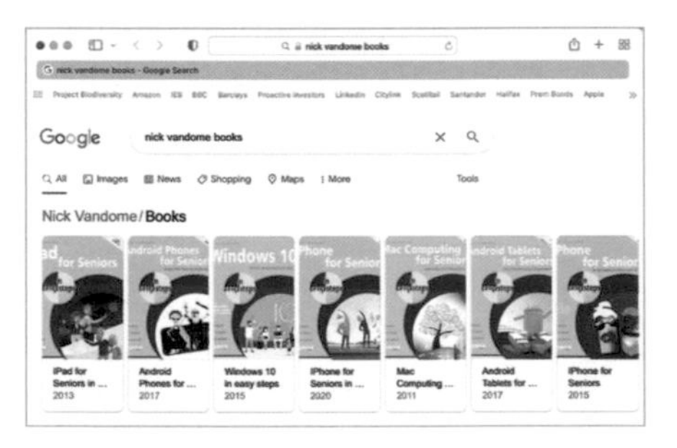

Click on a folder in Step 7 to open it directly in the Finder.

7 For other locations within your Mac, click on an item to see the details in the main window

For items found in the Mail app, click on an email to view its details in the main window

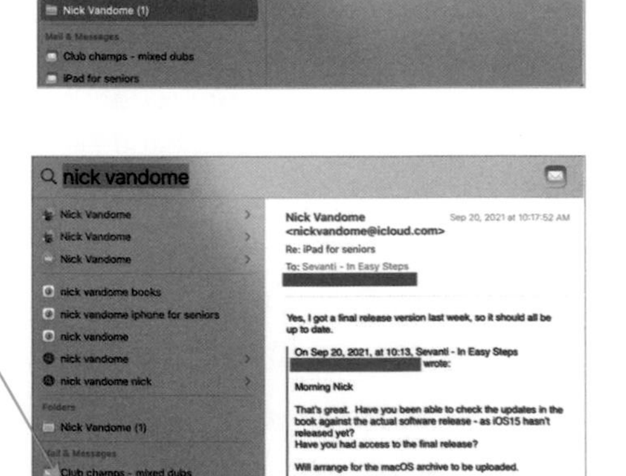

Finder search

This is the Search box in the top right-hand corner of the Finder and can be used to search for items within it. See page 69 for details about using Finder search.

Using Siri

Siri is the digital voice assistant that can be used to vocally search for a wide range of items from your Mac and the web. Also, the results can be managed in innovative ways so that keeping up-to-date is easier than ever.

Setting up Siri

To set up and start using Siri:

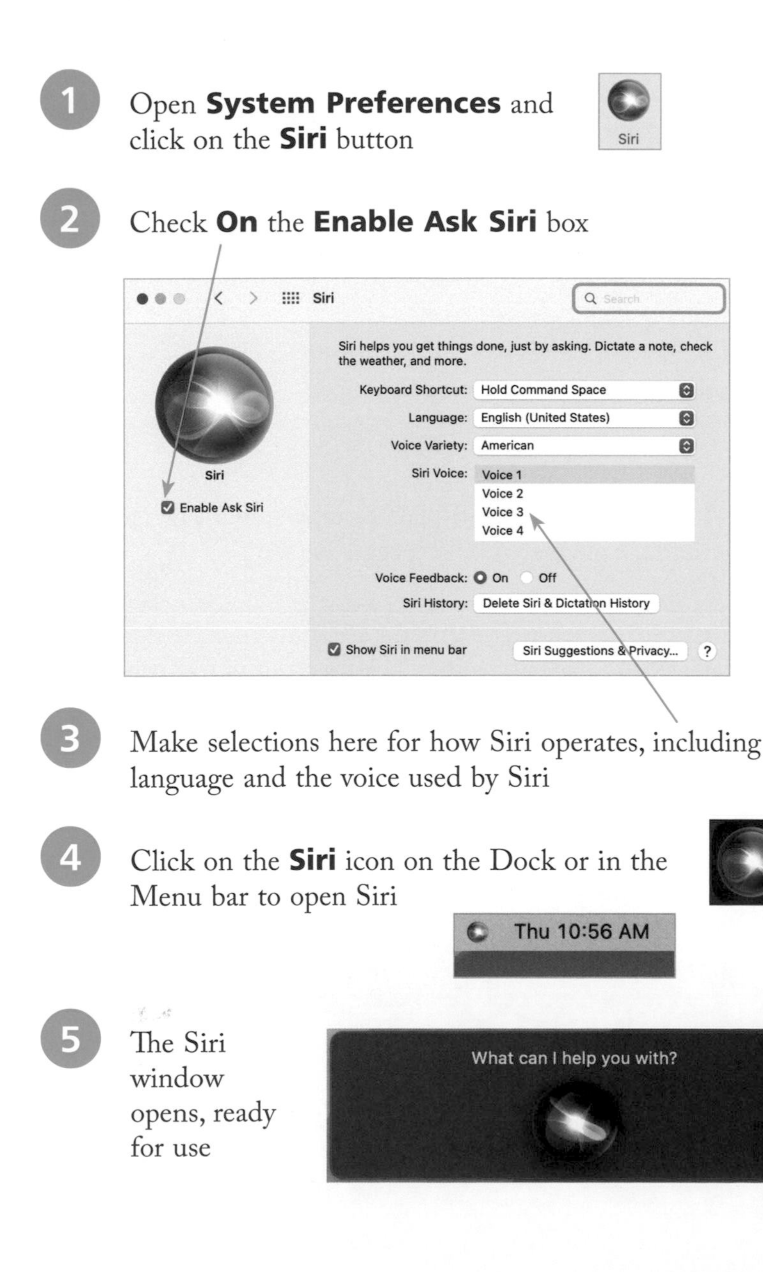

1. Open **System Preferences** and click on the **Siri** button
2. Check **On** the **Enable Ask Siri** box
3. Make selections here for how Siri operates, including language and the voice used by Siri
4. Click on the **Siri** icon on the Dock or in the Menu bar to open Siri
5. The Siri window opens, ready for use

Check **On** the **Show Siri in menu bar** box to show the Siri icon in the top right-hand corner of the main Apple Menu bar.

Show Siri in menu bar

An internal or external microphone is required in order to ask Siri questions.

Searching with Siri

Siri can respond to most requests, including playing a music track, checking the weather, opening photos, and adding calendar events. If you open Siri but do not ask a question, it will prompt you with a list of possible queries. Click on the microphone icon at the bottom of the panel to ask a question. It is also possible to ask questions of Siri while you are working on another document.

Siri can search for a vast range of items, from your Mac or the web. Some useful Siri functions include:

Hot tip

It is possible to ask Siri to search for specific types of documents; e.g. those created in Pages, or PDF files.

Searching for documents

This can be refined by specifying documents covering a specific date or subject.

1. Ask Siri to find and display documents with certain criteria; e.g. created by a specific person

2. Ask Siri to refine the criteria; e.g. only include documents created in a specific app. Click on an item to open it in its related app

Opening apps

Apps can be opened by Siri, with a simple "open" command:

Working with apps

Siri can also be used to add content to certain apps, such as adding an event to the calendar, adding a reminder, or creating a new note and adding content to it. This is done by asking Siri to create a new note, and you will then be prompted to add text for the note (you do not have to say "Siri" again before adding content to the note).

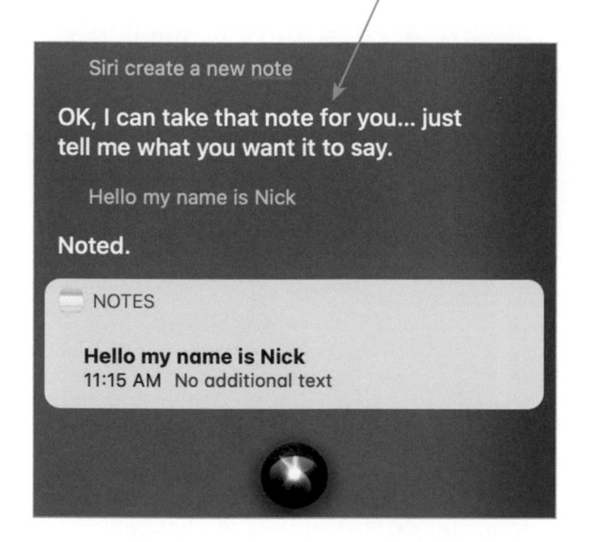

Move the cursor over the Siri search results and click on the cross in the top left-hand corner of the Siri window to close it. Otherwise, it can be left open for more queries and searches.

Multitasking

The Siri search window can be left open on the Desktop so that you can ask Siri a question while you are working on something else; e.g. while you are working on a document, ask Siri to display a relevant web page.

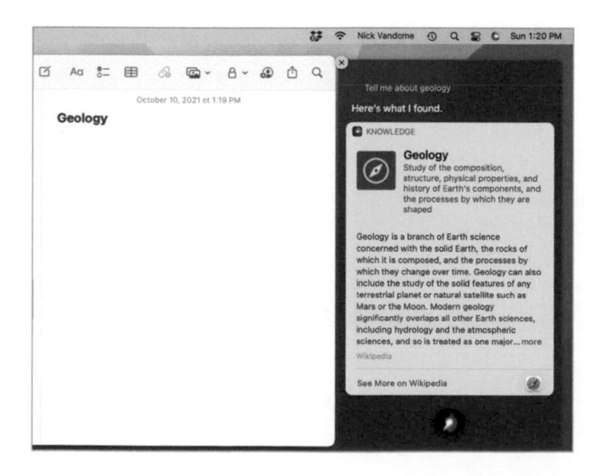

Accessibility

In all areas of computing it is important to give as many people as possible access to the system. This includes users with visual impairments and also people who have problems using the mouse and keyboard. In macOS this is achieved through the functions of the Accessibility section of System Preferences. To use these:

Another option in the Accessibility window is for **Switch Control**, which enables a Mac to be controlled by a variety of devices including the mouse, keypad and gamepad devices.

Experiment with the VoiceOver function, if only to see how it operates. This will give you a better idea of how visually-impaired users access information on a computer.

1. Click on the **Accessibility** button in the **System Preferences** folder to display the accessibility options in the left-hand panel

2. Click on the **Display** option to change the settings for display colors and contrast, and to increase the cursor size

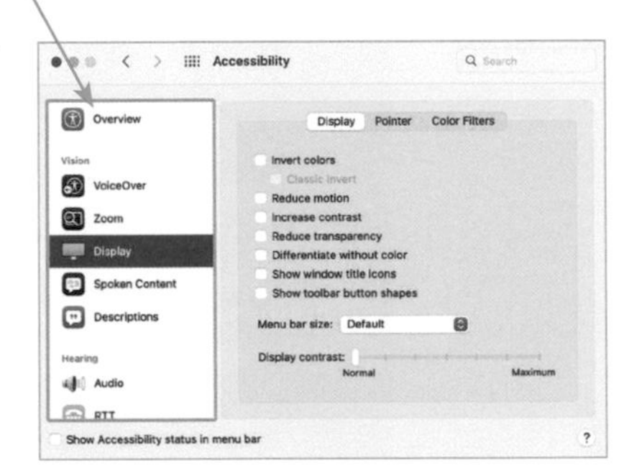

3. Click on the **Zoom** option to change the settings for zooming in on the screen

4. Click on the **VoiceOver** option to enable VoiceOver, which provides a spoken description of what is on the screen

5. Click on the **Audio** option to select an on-screen flash for alerts, and settings for how sound is played

6. Click on the **Keyboard** option to access settings for customizing the keyboard

7. Click on the **Pointer Control** option to access settings for customizing the mouse and trackpad

8. Click on the **Dictation** option to select preferences for using spoken commands

The Spoken Word

macOS Monterey not only has numerous apps for adding text to documents, emails and messages; it also has a Dictation function so that you can speak what you want to appear on screen. To set up and use the Dictation feature:

1. Click on the **Dictation** tab in **System Preferences** > **Keyboard**

2. By default, the Dictation option is **Off**

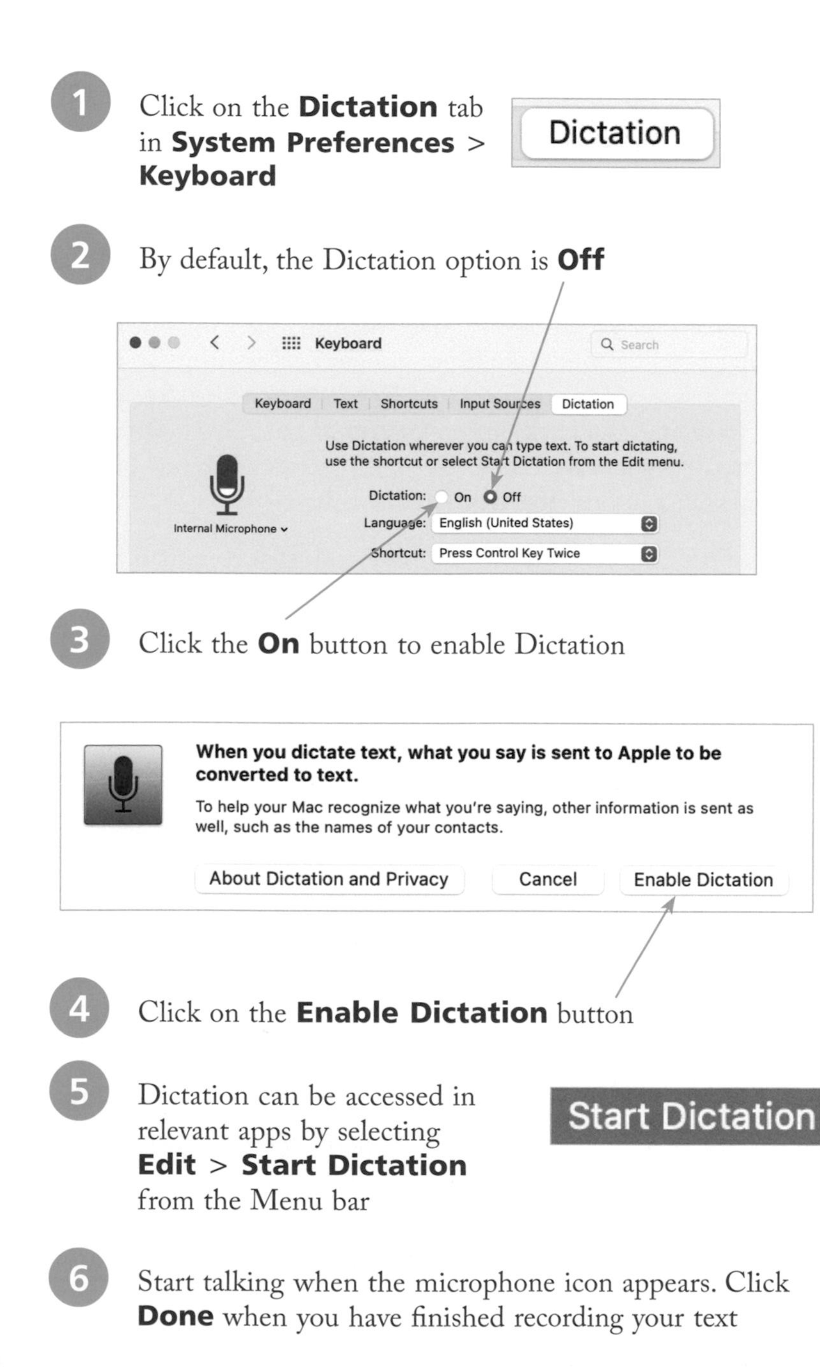

3. Click the **On** button to enable Dictation

4. Click on the **Enable Dictation** button

5. Dictation can be accessed in relevant apps by selecting **Edit** > **Start Dictation** from the Menu bar

6. Start talking when the microphone icon appears. Click **Done** when you have finished recording your text

Punctuation can be added with the Dictation function, by speaking commands such as "comma" or "question mark". These will then be converted into the appropriate symbols.

Dictation settings can also be found in **System Preferences** > **Accessibility** – see Step 8 on the previous page.

Shutting Down

The Apple menu (which can be accessed by clicking on the Apple icon in the top-left corner of the Desktop, or any subsequent macOS window) is standardized in macOS. This means that it has the same options regardless of the app in which you are working. This has a number of advantages, not least being the fact that it makes it easier to shut down your Mac. When shutting down, there are three options that can be selected:

- **Sleep**. This puts the Mac into hibernation mode; i.e. the screen goes blank and the hard drive becomes inactive. This state is maintained until the mouse is moved or a key is pressed on the keyboard. This then wakes up the Mac and it is ready to continue work. It is a good idea to add a login password for accessing the Mac when it wakes up, otherwise other people could wake it up and gain access to it. To add a login password, go to **System Preferences** > **Security & Privacy** and click on the **General** tab. Check **On** the **Require password** checkbox and select from one of the timescale options (**immediately** is best).
- **Restart**. This closes down the Mac and then restarts it again. This can be useful if you have added new software and your computer requires a restart to make it active.
- **Shut Down**. This closes down the Mac completely once you have finished working.

Click here to access the **Apple menu**

Click here to access one of the shut-down options

Finder File Edit View Go
About This Mac
System Preferences...
App Store... 9 updates
Recent Items >
Force Quit... ⌥⌘⎋
Sleep
Restart...
Shut Down...
Lock Screen ^⌘Q
Log Out Nick Vandome... ⇧⌘Q

When shutting down, make sure you have saved all of your open documents, although macOS will prompt you to do this if you have forgotten.

If a password is added, the Lock screen will appear whenever the Mac is woken from sleep and the password will be required to unlock it.

Users of the Apple Watch can unlock their Mac running macOS Monterey, without a password.

macOS Monterey has a **Resume** feature, where the Mac opens up in the same state as when it was shut down. See page 54 for details.

2 Getting Up and Running

This chapter looks at some of the essential features of macOS Monterey. These include the Dock for organizing and accessing all of the elements of your Mac computer; System Preferences for the way your Mac looks and operates; and items for arranging folders and files. It also details the online service – iCloud – for sharing digital content, including the Family Sharing feature for sharing with family members.

Introducing the Dock

The Dock is one of the main organizational elements of macOS. Its main function is to help organize and access apps, folders and files. In addition, with its background and icons, it also makes an aesthetically-pleasing addition to the Desktop. The main things to remember about the Dock are:

- It is divided into two: apps go on the left of the dividing line; all other items, and open apps, go on the right.

- It can be customized in several different ways.

By default, the Dock appears at the bottom of the screen:

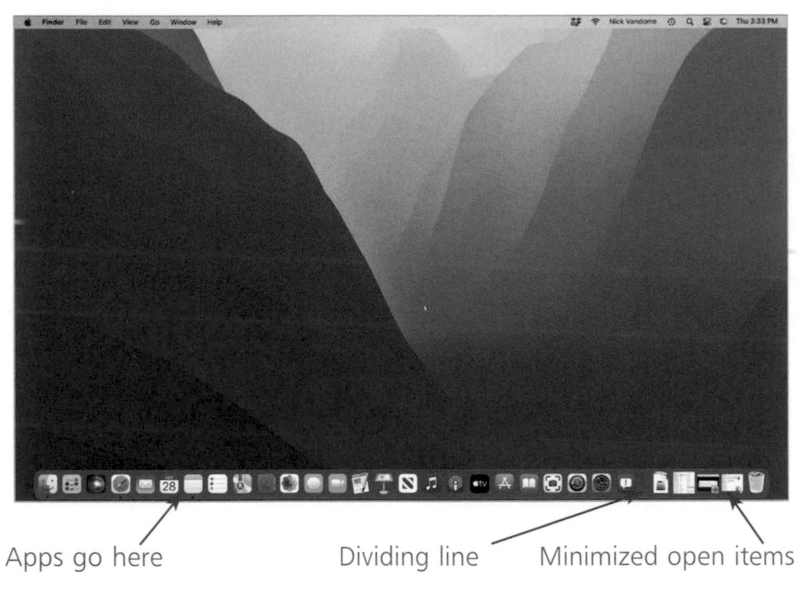

By default, the two icons to the right of the dividing line are:

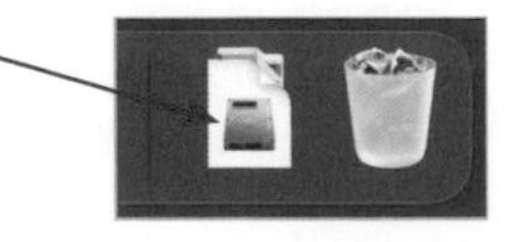

Downloads. This is for items that you have downloaded from the web. The items can be accessed from here, and opened or run.

Trash. This is where items can be dragged to if you want to remove them. It can also be used to eject removable devices, such as flashdrives, by dragging the device's icon over the Trash. It cannot be removed from the Dock.

The Dock is always displayed as a line of icons, but this can be orientated either vertically or horizontally.

The Downloads icon can be displayed as a folder or a Stack. To set this, **Ctrl** + **click** on the **Downloads** icon and select either **Folder** or **Stack** under the **Display as** option.

Apps on the Dock

Opening apps

When you open apps they appear on the Dock and can be worked with within the Dock environment.

Items on the Dock can be opened by clicking on them once, rather than having to double-click them. Once they have been accessed, the icon bobs up and down until the item is available.

1. Click once on an app to open it (either on the Dock, in the Finder Applications folder or the Launchpad). Once an app has been opened, it is displayed on the Dock with a black dot underneath it

2. When an open app is minimized it is displayed to the right of the dividing line

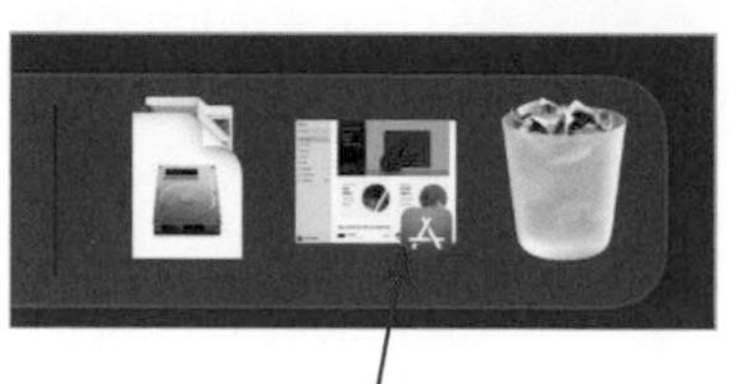

3. Click on an icon to the right of the dividing line to maximize it: it disappears from the Dock and displays at full size

4. If a window is minimized by clicking on this button, it goes back to the right-hand side of the dividing line on the Dock

5. Press and hold underneath an open app to view the available windows for the app (this will differ for individual apps, as some operate by using a single window)

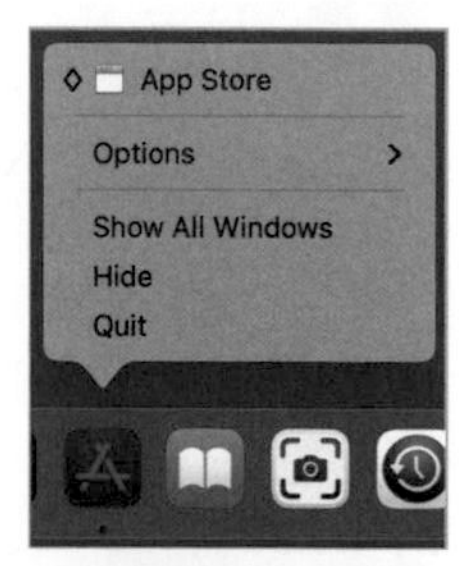

6. To close an open app, press and hold on the black dot underneath its icon on the Dock and click on the **Quit** button (or select its name on the Menu bar and select **Quit**)

Apps can be opened from the Dock, the Finder or the Launchpad. The Finder is covered in detail in Chapter 3, and see pages 90-91 for more on the Launchpad.

Some apps (such as Notes, Reminders and Calendar) will close when the active window is closed. Others (such as Pages, Keynote and Numbers) will remain open even if all of the windows are closed: the recently-accessed documents will be displayed as in the context menu in Step 5.

Setting Dock Preferences

As with most elements of macOS, the Dock can be modified in numerous ways. This can affect both the appearance of the Dock and the way it operates. To set Dock preferences:

Select **System Preferences** > **Dock & Menu Bar**

The Dock preferences allow you to change its size, orientation, the way icons appear with magnification, and effects for when items are minimized

The Dock cannot be moved by dragging it physically; this can only be done in the Dock preferences window.

Drag the Dock **Size:** slider to increase or decrease the size of the Dock

You cannot make the Dock size so large that some of the icons would not be visible on the Desktop. By default, the Dock is resized so that everything is always visible.

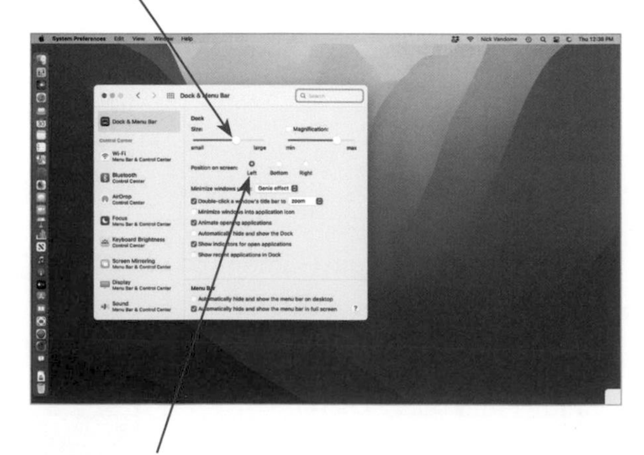

The **Position on screen:** options enable you to place the Dock on the left, right or bottom of the screen.

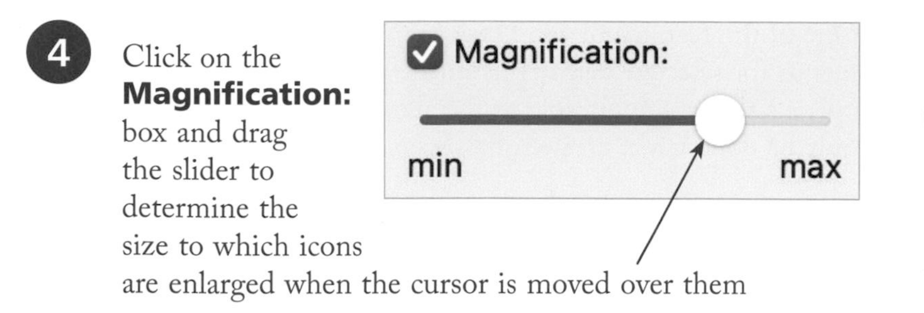

4. Click on the **Magnification:** box and drag the slider to determine the size to which icons are enlarged when the cursor is moved over them

Move the cursor over an icon on the Dock to see the magnification effect.

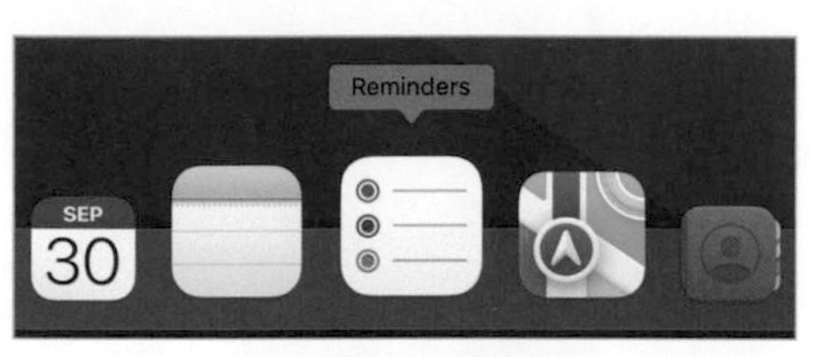

5. The **Genie effect** under the **Minimize windows using** option shrinks the item to be minimized, like a genie going back into its lamp

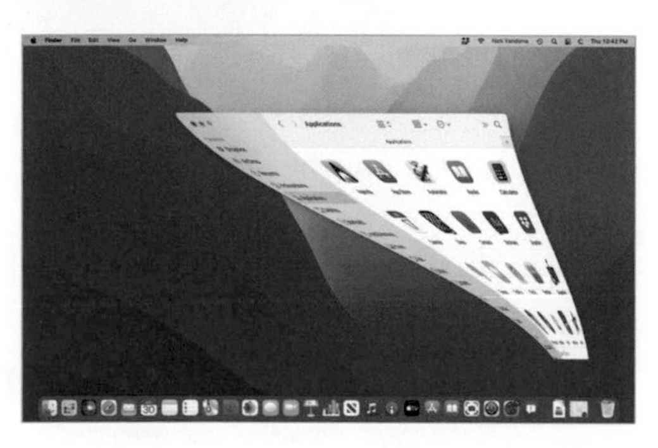

Manual resizing

In addition to changing the size of the Dock by using the Dock preferences dialog box, it can also be resized manually.

1. Drag vertically on the Dock dividing line to increase or decrease its size

Open windows can also be maximized and minimized by double-clicking on the Title bar (the area at the top of the window, next to the three window buttons).

Folders of documents can be added to the Dock. These are known as Stacks and are created by dragging a folder from a Finder window onto the right-hand side of the Dock. The files can be accessed by clicking on the Stack. Move the cursor over a Stack, then press **Ctrl** + **click** to access options for how the Stack is displayed. Stacks can also be created on the Desktop, by dragging a folder there and selecting **View** > **Stacks** from the Menu bar.

Dock Menus

One of the features of the Dock is that it can display contextual menus for selected items. This means that it shows menus with options that are applicable to the item being accessed. This can only be done when an item has been opened.

1. Click and hold on the black dot underneath an item's icon to display the item's individual context menu

2. Click on **Show in Finder** to see where the item is located on your computer

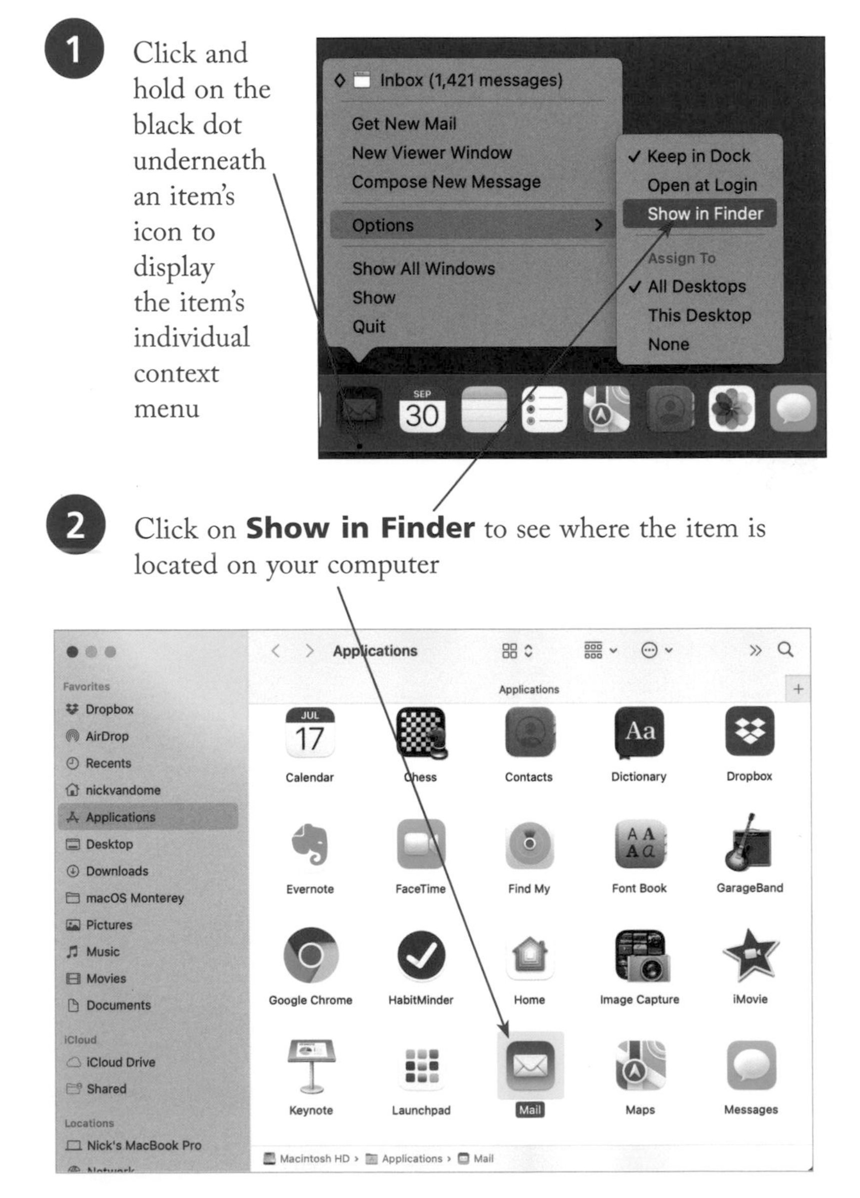

Click on **Quit** on the icon's contextual menu to close an open app or file, depending on which side of the dividing bar the item is located.

Working with Dock Items

Adding items

Numerous items can be added to the Dock; the only restriction is the size of monitor on which to display all of the Dock items. (The size of the Dock can be reduced to accommodate more icons, but you have to be careful that all of the icons are still legible.) To add items to the Dock:

Locate the required item in the Finder and drag it onto the Dock. All of the other icons move along to make space for the new one

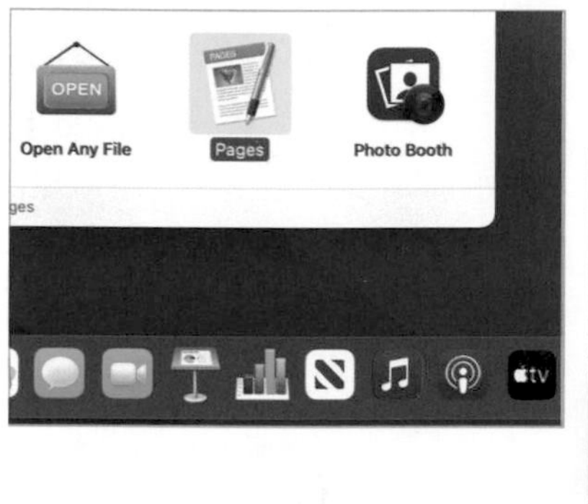

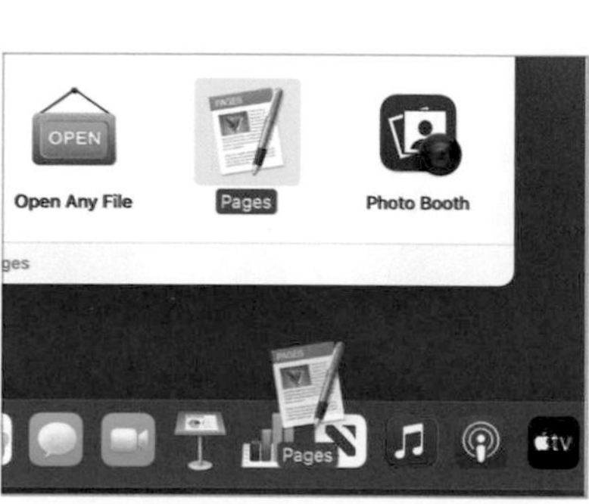

Keep in Dock

Every time you open a new app, its icon will appear in the Dock for the duration that the app is open, even if it has not previously been put in the Dock. If you then decide that you would like to keep it in the Dock, you can enable this as follows:

Click and hold on the black dot underneath an open app's icon

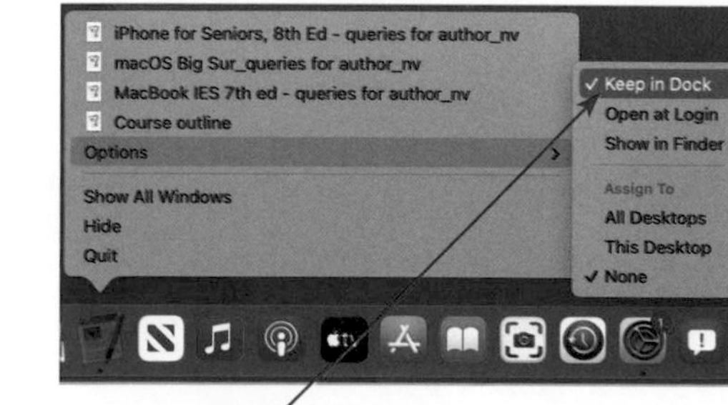

Select **Options** > **Keep in Dock** to ensure the app remains in the Dock when it is closed

If you are upgrading to macOS Monterey (rather than buying a new Mac), the Dock setup will be the same as the one used with the previous version of the operating system.

Icons on the Dock are shortcuts to the related item rather than the item itself, which remains in its original location.

You can add numerous items to the Dock, but it will automatically shrink to display all of its items if it becomes too big for the available space.

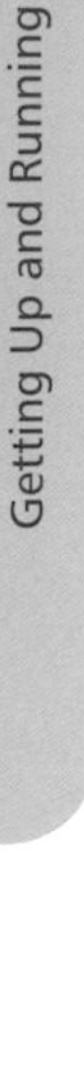

...cont'd

Removing items

Any item on the left of the dividing line, except the Finder, can be removed from the Dock. However, this does not remove it from your computer; it just removes the shortcut for accessing it. You will still be able to locate it in its folder in the Finder and, if required, drag it back onto the Dock. To remove items from the Dock:

When an icon is dragged from the Dock, it has to be moved a reasonable distance before the **Remove** alert appears.

1. Drag the item away from the Dock until it displays the **Remove** tag. The item disappears once the cursor is released. All of the other icons then move up to fill in the space

Removing open apps

You can remove an app from the Dock, even if it is open and running. To do this:

1. Drag an app off the Dock while it is running. Initially, the icon will remain on the Dock because the app is still open

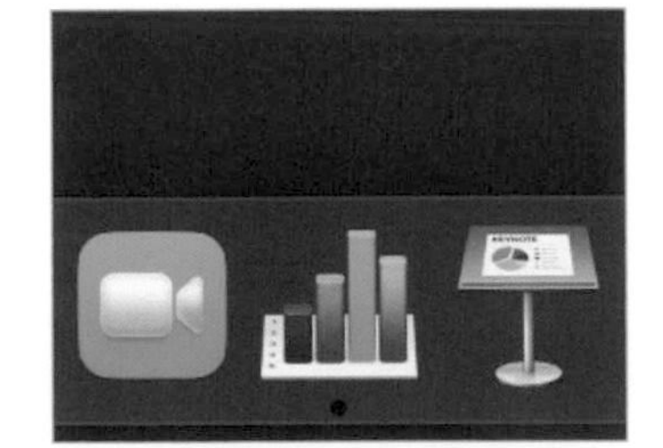

If **Keep in Dock** has been selected for an item (see page 31) the app will remain in the Dock even when it has been closed.

2. When the app is closed (click and hold on the dot underneath the item and select **Quit**) its icon will be removed from the Dock

Trash

The Trash folder is a location for placing items that you do not need anymore. However, when items are placed in the Trash, they are not removed from your computer. This requires another command, as the Trash is really a holding area before you decide you want to remove items permanently. The Trash can also be used for ejecting removable disks attached to your Mac.

Items can also be sent to the Trash by selecting them in the Finder and then selecting **File** > **Move to Trash** from the Menu bar.

Sending items to the Trash

Items can be sent to the Trash by dragging them from the location in which they are stored.

Drag an item over the **Trash** icon to place it in the Trash folder

Click once on the **Trash** icon on the Dock to view its contents

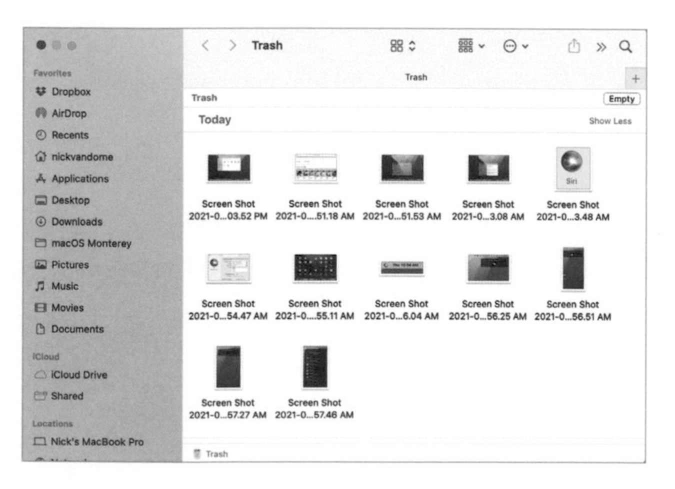

All of the items within the Trash can be removed in a single command: select **Finder** > **Empty Trash** from the Menu bar to remove all of the items in the Trash folder.

Using System Preferences

In macOS there are preferences that can be set for just about every aspect of your computer. This gives you greater control over how the interface looks and how the operating system functions. To access System Preferences:

1. Click on this icon on the Dock or from the Launchpad

Apple ID, iCloud, Media & App Store. This contains options for setting up an Apple ID for use with iCloud, the online service.

Family Sharing. This can be used to set up and use Family Sharing, for sharing music, movies, games, apps and books with up to five other family members.

General. Options for the overall look of buttons, menus, windows and scroll bars.

Desktop & Screen Saver. This can be used to change the Desktop background and the screen saver.

Dock & Menu Bar. Options for the way the Dock and the top Menu bar look and operate.

Mission Control. This gives you a variety of options for managing all of your open windows and apps.

Siri. This can be used to turn on or off the digital voice assistant, Siri, and also set voice style and language.

Spotlight. This can be used to specify settings for the macOS search facility, Spotlight.

Language & Region. Options for the language used on your Mac.

Notifications & Focus. This can be used to set up how you are notified about items such as email, messages and software updates.

Internet Accounts. This can be used to link to other online accounts that you have (see the third Hot tip).

Passwords. This can be used to manage, and change, passwords you have for accessing specific websites.

Users & Groups. This can be used to create accounts for different users on your Mac.

The General system preferences have an option for Dark Mode. If this is activated, the background is turned black, with the text inverted as white. This will apply to all compatible apps, once Dark Mode is activated. It can also be set to be turned on automatically, depending on the time of day.

The Desktop & Screen Saver system preferences have an option for Dynamic Desktop, which can be used to display different versions of the Desktop background, depending on the time of day.

The **Internet Accounts** section can be used to set up email accounts with webmail services such as Google.

Accessibility. This can be used to set options for users who have difficulty with sight, hearing or motor issues.

Screen Time. This can be used to show screen usage and limit access to the computer and various online functions.

Extensions. This can be used to customize your Mac with extensions and plug-ins from Apple and third-party developers.

Security & Privacy. This enables you to secure your Home folder with a master password, for added security.

Software Update. This can be used to specify how software updates are handled. It connects to the App Store for updates.

Network. This can be used to specify network settings for connecting to the internet or other computers.

Bluetooth. Options for attaching Bluetooth wireless devices.

Sound. Options for adding sound effects, and playing and recording sound.

Keyboard. Options for how the keyboard functions; your keyboard shortcuts; and dictation settings.

Trackpad. Options for when you are using a trackpad.

Mouse. Options for how the mouse functions.

Displays. Options for the screen display, such as resolution.

Printers & Scanners. Options for selecting printers and scanners.

Energy Saver. Options for managing the battery and energy settings.

Date & Time. Options for changing the computer's date and time to time zones around the world.

Sharing. This can be used to specify how files are shared over a network. This is also covered on page 169.

Time Machine. This can be used to configure and set up the macOS backup facility.

Startup Disk. This can be used to specify the disk from which your computer starts up. This is usually the macOS volume.

See pages 162-164 for more information about setting up and using Screen Time.

macOS Monterey supports multiple displays, which means you can connect your Mac to two or more displays and view different content on each one. The Dock appears on the active screen, and each screen also has its own Menu bar. Different full-screen apps can also be viewed on each screen.

Control Center

The Control Center has been a feature of Apple's mobile devices – the iPhone and iPad – for a number of years, and it is now available with macOS Monterey. It is a panel containing shortcuts to some of the most commonly-used options and functions within System Preferences.

Accessing the Control Center

The Control Center can be accessed from the top toolbar on the Desktop or from within any app. To do this:

Focus can be used to specify times during which notifications and FaceTime calls can be muted. Screen Mirroring can be used to display the Mac's screen on another screen.

1. To open the Control Center, click on this button on the top toolbar

2. Click on these buttons to access options for **Focus**, **Keyboard Brightness**, and **Screen Mirroring**

3. Click on these buttons to access options for turning Wi-Fi on or off; turning Bluetooth on or off; or specifying who can use AirDrop for sharing items

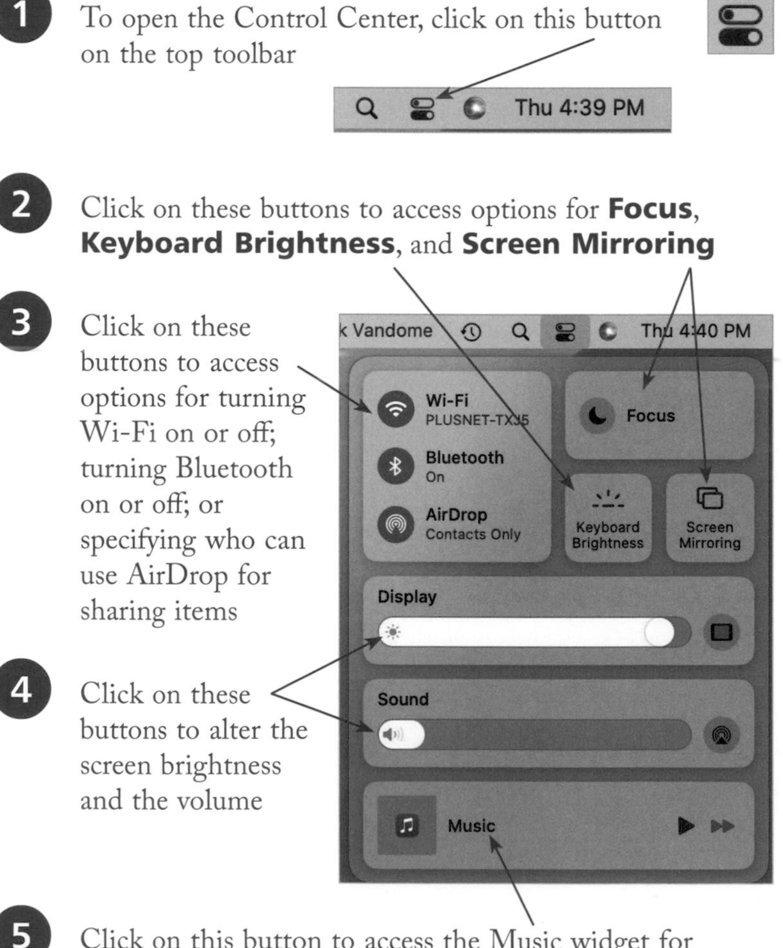

4. Click on these buttons to alter the screen brightness and the volume

5. Click on this button to access the Music widget for playing and controlling tracks from the Music app

AirDrop is the functionality for sharing items wirelessly between compatible devices. Click once on the **AirDrop** button in the Control Center and specify whether you want to share with **Contacts Only** or **Everyone**. Once AirDrop is set up, you can use the **Share** button in compatible apps to share items such as photos with any other AirDrop users in the vicinity.

Managing the Control Center

Items within the Control Center can be managed within System Preferences.

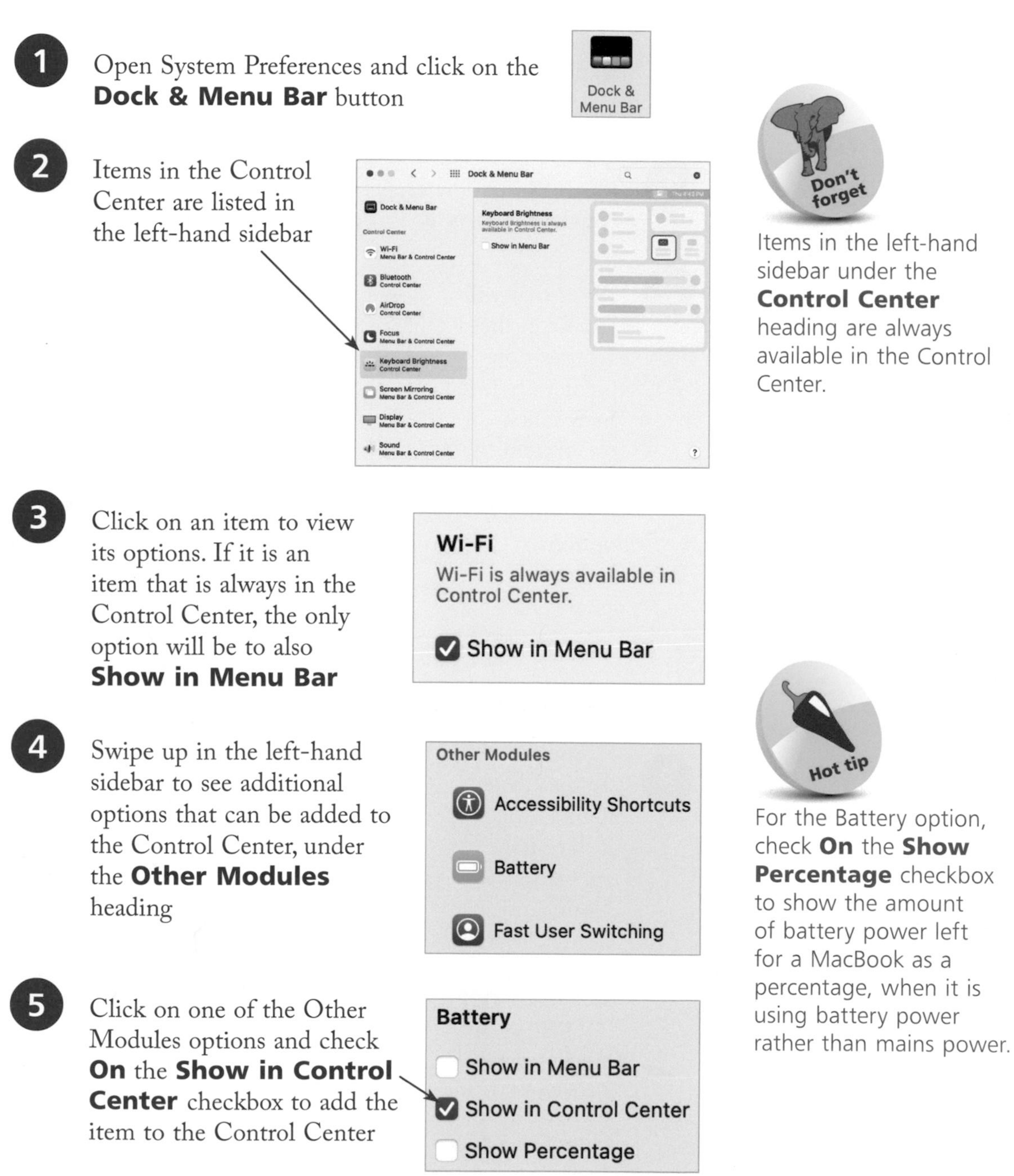

1. Open System Preferences and click on the **Dock & Menu Bar** button
2. Items in the Control Center are listed in the left-hand sidebar
3. Click on an item to view its options. If it is an item that is always in the Control Center, the only option will be to also **Show in Menu Bar**
4. Swipe up in the left-hand sidebar to see additional options that can be added to the Control Center, under the **Other Modules** heading
5. Click on one of the Other Modules options and check **On** the **Show in Control Center** checkbox to add the item to the Control Center

Don't forget

Items in the left-hand sidebar under the **Control Center** heading are always available in the Control Center.

Hot tip

For the Battery option, check **On** the **Show Percentage** checkbox to show the amount of battery power left for a MacBook as a percentage, when it is using battery power rather than mains power.

Notifications and Widgets

The Notification Center provides a single location to view all of your emails, messages, updates and alerts. Notifications appear in the top right-hand corner of the screen. In macOS Monterey, the Notification Center now includes widgets that can display a range of useful information. Widgets can be edited within the Notification Center itself, and apps that display notifications are set up within System Preferences. To do this:

The Notification Center has been updated in macOS Monterey, to include the Focus feature; see pages 42-43.

1. Open System Preferences and click on the **Notifications & Focus** button

Notifications can be accessed regardless of the app in which you are working, and they can be actioned directly without having to leave the active app.

2. Items that will appear in the Notification Center are listed here. Click on an item to select it and set its notification options

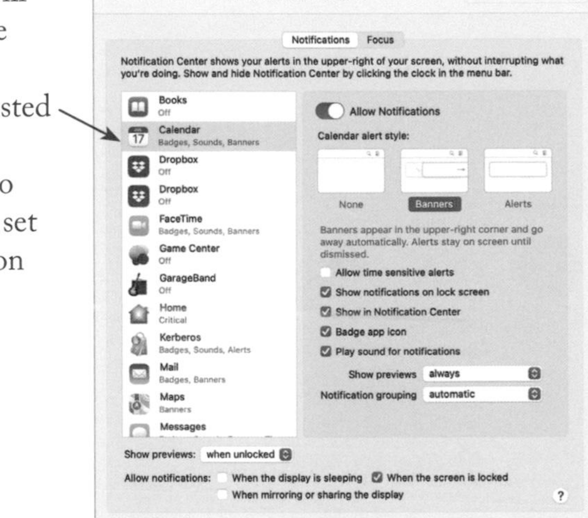

Twitter and Facebook feeds can also be set up to appear in the Notification Center, if you have accounts with these sites.

3. To disable an item so that it does not appear in the Notification Center, select it as above and check **Off** the **Show in Notification Center** box

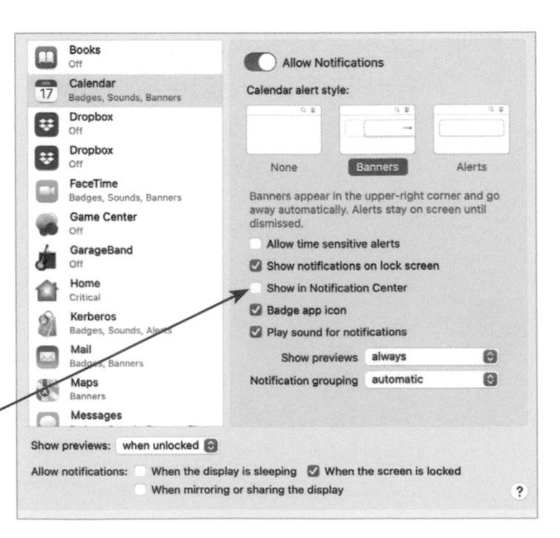

Viewing notifications

Notifications appear in the Notification Center. The way they appear can be determined in System Preferences.

1. Select an alert style. A banner alert comes up on the screen and then disappears after a few seconds

2. The **Alerts** option shows the notification, and it stays on screen until dismissed (such as this one for Reminders)

3. Click on this clock in the top right-hand corner of the screen to view all of the items in the Notification Center. Click on it again to hide the Notification Center

4. Notifications that have been set up in System Preferences appear at the top of the Notification Center. Widgets appear below any notifications

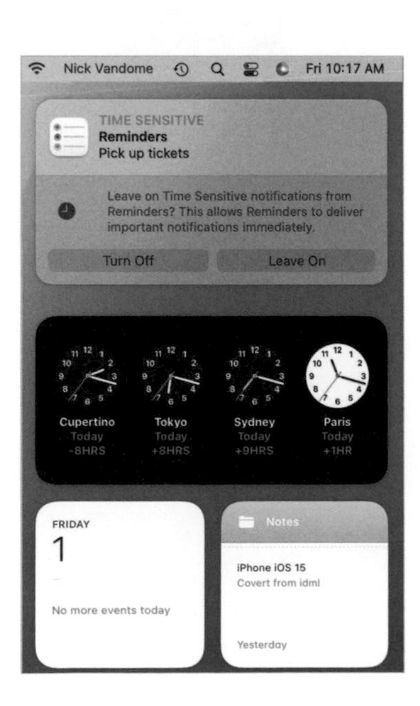

The Notification Center can also be displayed using a trackpad or Magic Trackpad by dragging with three fingers from right to left, starting from the far-right edge.

Software updates can also appear in the Notification Center, when they are available.

...cont'd

Using widgets

Widgets that appear in the Notification Center display can be managed and edited directly from the Notification Center panel. To do this:

Swipe down to the bottom of the Notification Center in Step 4 on page 39 and tap on the **Edit Widgets** button

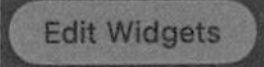

Existing widgets in the Notification Center are shown in the right-hand panel. The middle panel displays widgets that can be added to the Notification Center

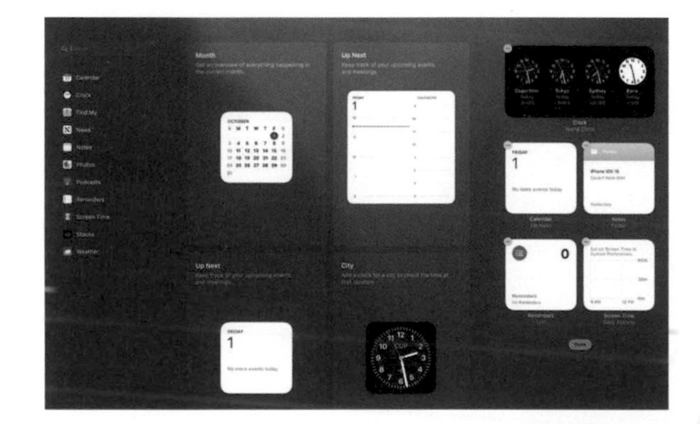

3. All of the available widgets are shown in the left-hand panel. Click on one to view details about it in the middle panel so that it can be added to the Notification Center, if required

Click on this button in the top left-hand corner of a widget in Step 2 to remove it from the Notification Center:

4. Click on a widget in the middle panel to add it to the Notification Center, or move the cursor over the widget and tap on the green **+** button

Several versions of the same widget can be added to the Notification Center, at different sizes, if required.

5. Some of the widgets have an option for adding them to the Notification Center at different sizes. If this is available for a widget these buttons will appear below the widget in the middle panel in Step 2 on the previous page. The options are for **Small**, **Medium** or **Large**

Some widgets can be edited to display different information – for instance, the Weather widget can display forecasts from numerous locations around the world. To see if a widget is editable, move the cursor over it in the right-hand panel in Step 2 on the previous page. If the widget is editable, click on the **Edit Widget** button.

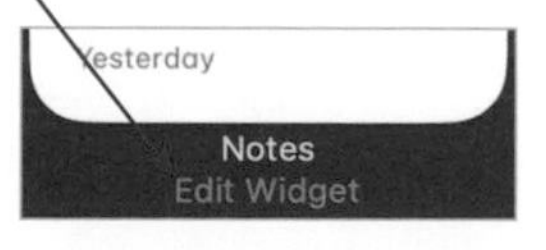

Focus

Another option for managing your notifications is the Focus function, which can be used to control when you are notified for certain items. The Focus function can be used to limit notifications when you are performing certain actions – e.g. reading or relaxing – and it can also specify certain people and apps that are allowed to contact you and send notifications. The Focus function can be set up within System Preferences, and also accessed from the Control Center. To do this:

Focus is a new feature in macOS Monterey.

The Focus feature contains Do Not Disturb, which was a stand-alone feature in previous versions of macOS Monterey. To set this up, click on the **Do Not Disturb** button in Step 2 and select the required options, and times, for when you want Do Not Disturb to operate.

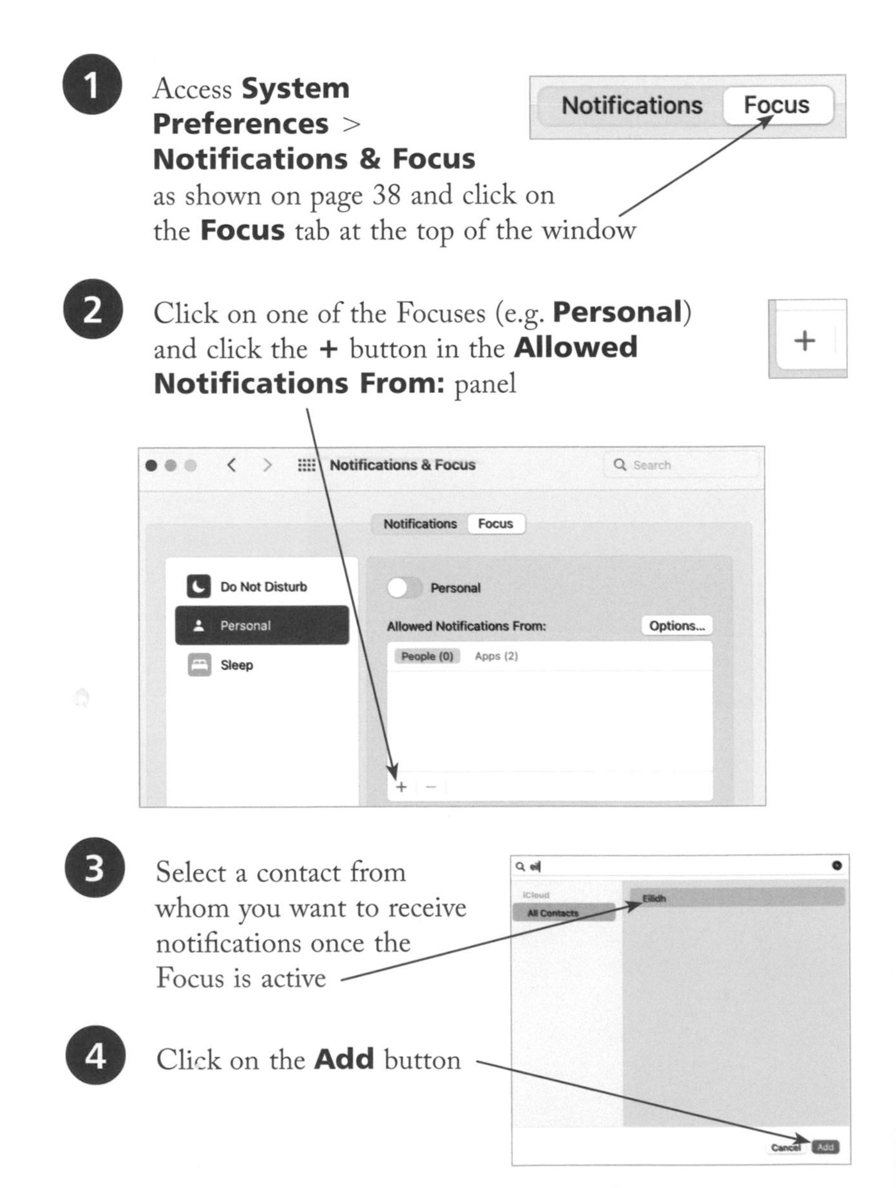

1. Access **System Preferences** > **Notifications & Focus** as shown on page 38 and click on the **Focus** tab at the top of the window

2. Click on one of the Focuses (e.g. **Personal**) and click the **+** button in the **Allowed Notifications From:** panel

3. Select a contact from whom you want to receive notifications once the Focus is active

4. Click on the **Add** button

Different Focuses can be set for specific activities, and the settings for the Focus can be applied according to the required activity.

5. The selected person is added to your **Allowed Notifications From:** list

6. Click on the **Apps** tab and repeat the process for adding apps to be allowed when the Focus is active

7. Drag the **Personal** button **On** to make the Focus active

8. Click on the **+** button in the bottom left-hand corner of the main **Notifications & Focus** window to add more Focuses to the left-hand panel in Step 2 on the previous page

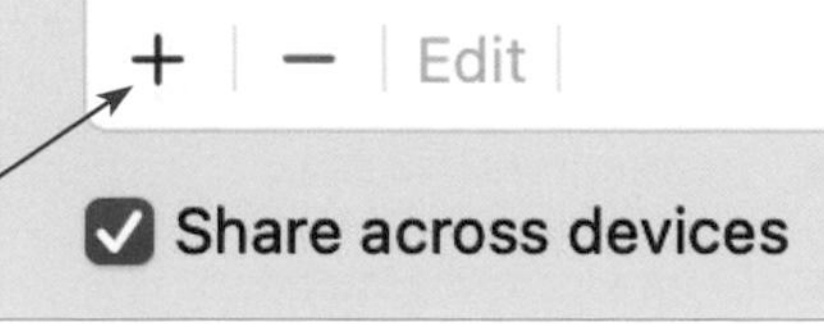

Beware

If you add too many people in Step 5, this could negate the effect of using the specific Focus.

Once a Focus has been created, access the Control Center as shown on page 36 and click on the **Focus** button to add more Focuses, and turn existing ones **On** or **Off**, and view their details.

Check **On** the **Share across devices** box in Step 8 to apply the Focus to all of your Apple devices sharing the same Apple ID.

Using iCloud

Cloud computing is an attractive proposition and one that has gained great popularity in recent years. As a concept, it consists of storing your content on an external computer server. This not only gives you added security in terms of backing up your information; it also means that the content can then be shared over a variety of devices.

The standard iCloud service is free, and this includes an iCloud email address and 5GB of online storage. (*Correct at the time of printing.*)

iCloud is Apple's consumer cloud-computing product that consists of online services such as email; a calendar; notes; contacts; and saving photos and documents. iCloud provides users with a way to save and back up their files and content to the online service, and then use them across their Apple devices such as other Mac computers, iPhones, iPads and iPod Touches.

About iCloud

Once iCloud has been set up (see the next page) it can be accessed from within System Preferences.

Click on the **Apple ID** button

Click on the **iCloud** button

You can use iCloud to save and share the following between your different devices, with an Apple ID account (see the next page):

There is also a version of iCloud for Windows, which can be accessed for download from the Apple website at **https://support.apple.com/en-us/HT204283**

- Photos
- Mail and Safari
- Documents
- Backups
- Notes
- Reminders
- Contacts and Calendar

When you save an item to iCloud it automatically pushes it to all of your other compatible devices; you do not have to manually sync anything as iCloud does it all for you.

Setting up iCloud

To use iCloud with macOS Monterey you need to first have an Apple ID. This is a service you can register for to be able to access a range of Apple facilities, including iCloud. You can register with an email address and a password. When you first start using iCloud you will be prompted for your Apple ID details. If you do not have an Apple ID you can create one at this point.

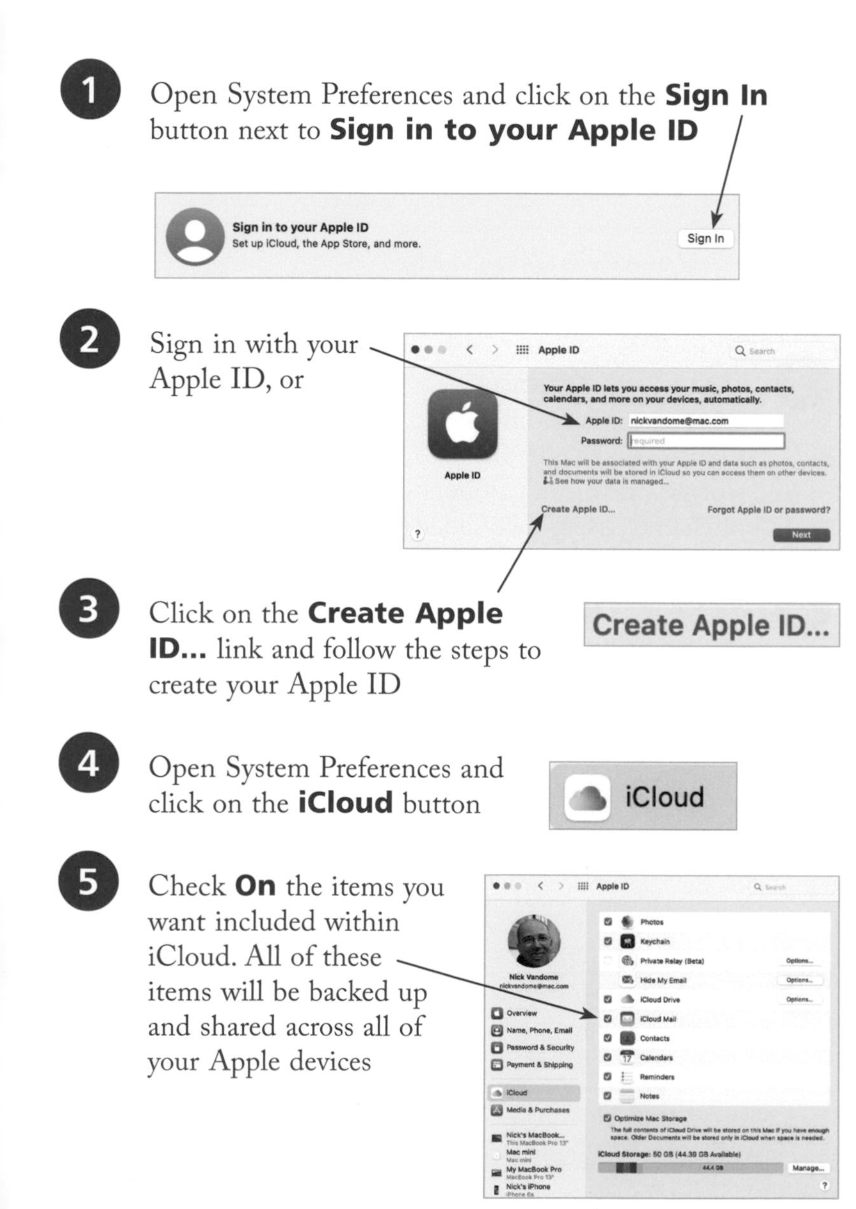

1. Open System Preferences and click on the **Sign In** button next to **Sign in to your Apple ID**
2. Sign in with your Apple ID, or
3. Click on the **Create Apple ID...** link and follow the steps to create your Apple ID
4. Open System Preferences and click on the **iCloud** button
5. Check **On** the items you want included within iCloud. All of these items will be backed up and shared across all of your Apple devices

When you have an Apple ID and an iCloud account, you can also use the iCloud website to access your content. Access the website at **www.icloud.com** and log in with your Apple ID details.

The online iCloud service includes your online email service; Contacts; Calendars; Reminders; Notes; and versions of Pages, Keynote and Numbers. You can log in to your iCloud account from any internet-enabled device.

The iCloud Drive option can be used to store items created in the productivity apps Pages, Numbers, and Keynote. Documents are stored in the Files app and they can be stored and accessed here and from iCloud.

Upgrading to iCloud+

iCloud+ is an enhancement to iCloud that enables you to add more storage to your iCloud account and also access new security options. To subscribe to iCloud+:

iCloud+ is a new feature in macOS Monterey.

Click on the **iCloud** button as shown in Step 4 on page 45

The amount of storage that has been used is indicated by the colored bar at the bottom of the window (e.g. yellow for photos, and blue for email). Click on the **Manage...** button

iCloud Storage: 5 GB (0.5 GB Available)
Manage...

iCloud+ is the equivalent of the iCloud upgrade service, from the free version, in previous versions of macOS Monterey. iCloud+ also has a few additional features; see the next page for details.

Click on the **Change Storage Plan...** button to increase the amount of storage (the default amount is 5GB, which is provided free of charge), using the iCloud+ service

Change Storage Plan...

Select an iCloud+ storage plan and click on the **Next** button to confirm the selection

Upgrade iCloud+
iCloud+
50 GB £0.79 per month Renews 23/10/21
Upgrade Options
200 GB £2.49 per month Can be shared with your family
2 TB £6.99 per month Can be shared with your family
Prices include VAT (where applicable)
Your storage plan will automatically renew. You can cancel at any time. Learn more
Downgrade Options... Cancel Next

Prices are shown in local currencies.

The iCloud storage is increased, according to the new iCloud+ storage plan

iCloud Storage: 50 GB (44.41 GB Available)
44.4 GB
Manage...

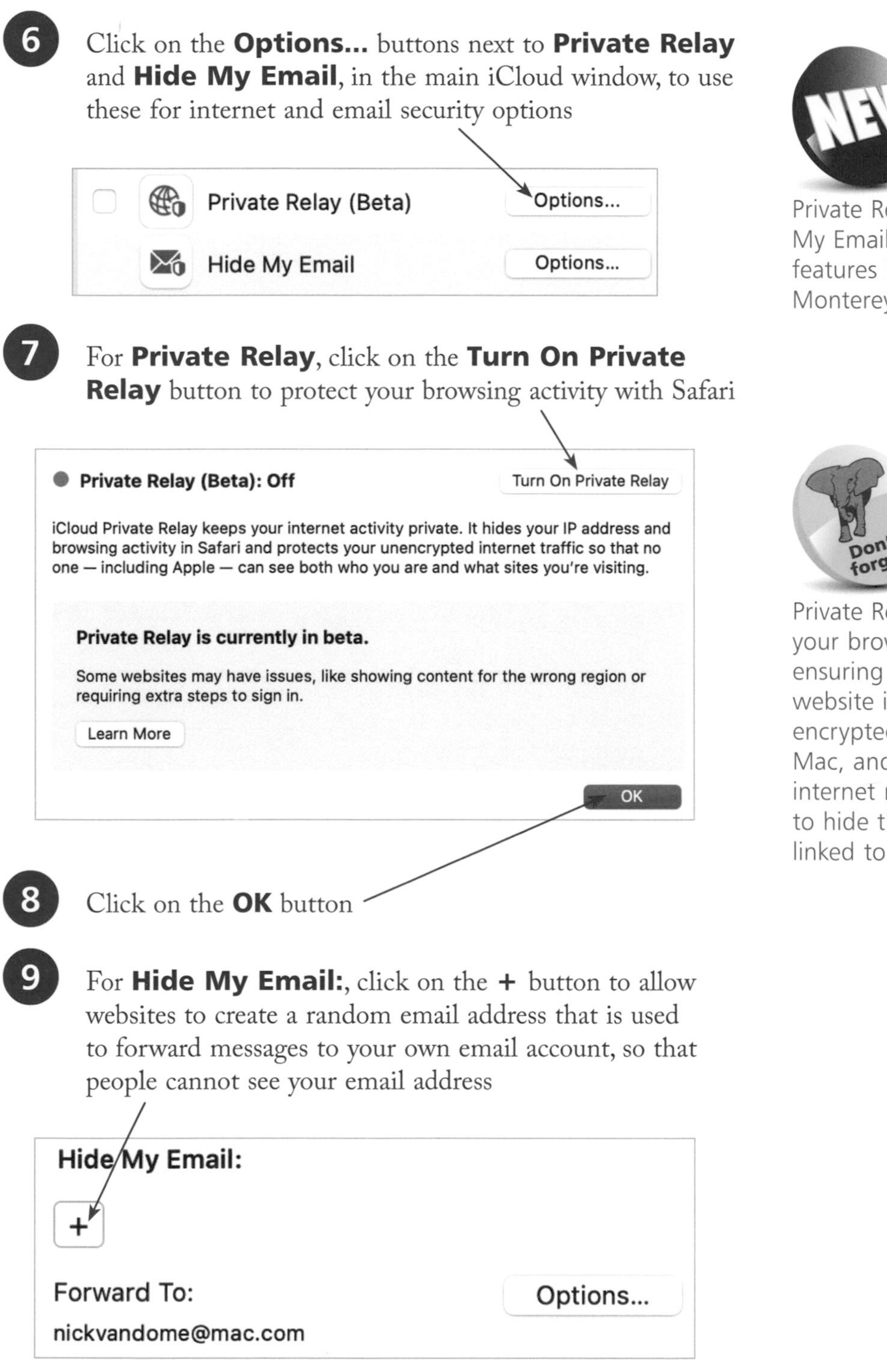

Private Relay (Beta)
Options...
Hide My Email
Options...
Private Relay (Beta): Off
Turn On Private Relay
iCloud Private Relay keeps your internet activity private. It hides your IP address and browsing activity in Safari and protects your unencrypted internet traffic so that no one — including Apple — can see both who you are and what sites you're visiting.
Private Relay is currently in beta.
Some websites may have issues, like showing content for the wrong region or requiring extra steps to sign in.
Learn More
OK
Hide My Email:
+
Forward To:
Options...
nickvandome@mac.com

6. Click on the **Options...** buttons next to **Private Relay** and **Hide My Email**, in the main iCloud window, to use these for internet and email security options

7. For **Private Relay**, click on the **Turn On Private Relay** button to protect your browsing activity with Safari

8. Click on the **OK** button

9. For **Hide My Email:**, click on the **+** button to allow websites to create a random email address that is used to forward messages to your own email account, so that people cannot see your email address

NEW

Private Relay and Hide My Email are new features in macOS Monterey.

Private Relay protects your browsing by ensuring that any website interactions are encrypted from your Mac, and two separate internet relays are used to hide the IP address linked to your Mac.

About Family Sharing

Being able to share digital content with other people is an important part of the digital world, and with macOS Monterey and iCloud, the Family Sharing function enables you to share items that you have downloaded from the relevant online Apple store with your Apple ID, such as music and movies, with up to five other family members, as long as they have an Apple ID. To set up Family Sharing:

Once Family Sharing has been set up, items that can be shared, including subscriptions and new purchases, will be listed in the left-hand sidebar in Step 2. Click on an item to access it: once this has been done it will be available to other Family Sharing members.

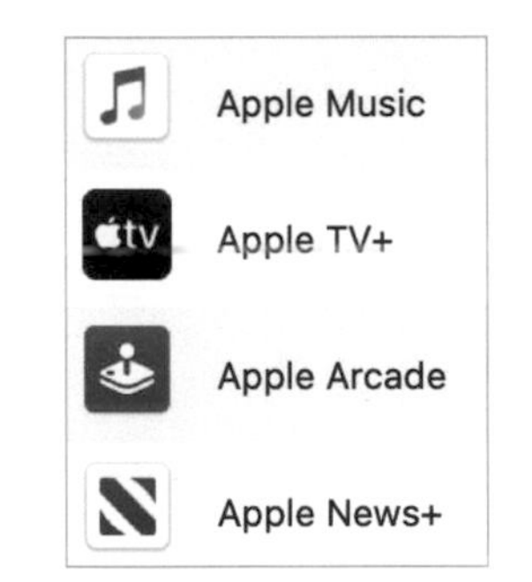

1. Open System Preferences and click on the **Family Sharing** button

2. Click on the **Get Started** button (the initial setup for Family Sharing involves a step-by-step process, during which you can include a payment method for Family Sharing items)

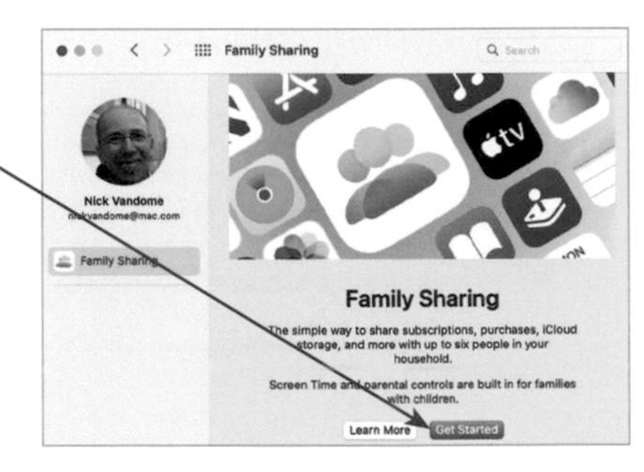

3. Click on the **Invite People** button

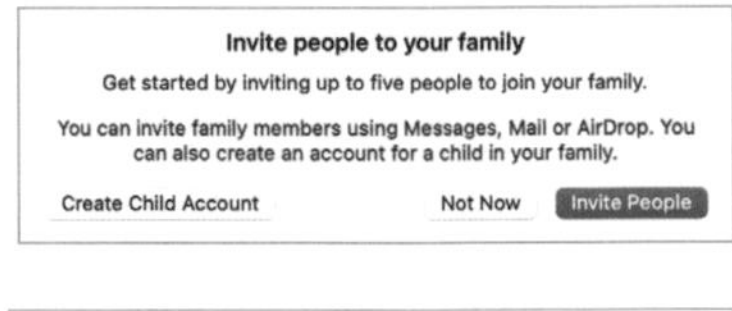

4. Select an option for sending the invitation and click on the **Continue** button

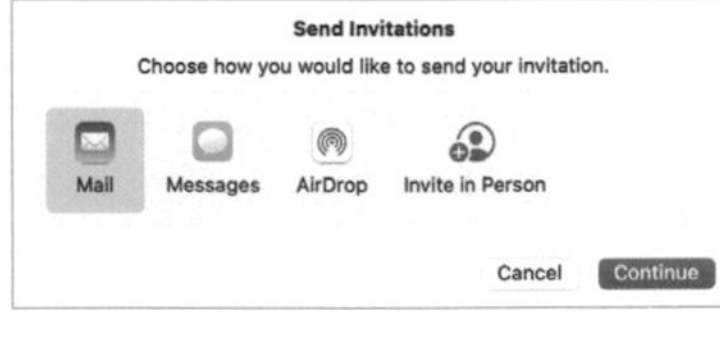

5. Select the person who you want to invite, add text as required, and send the invite. The recipient has to accept the invite to join your Family Sharing group

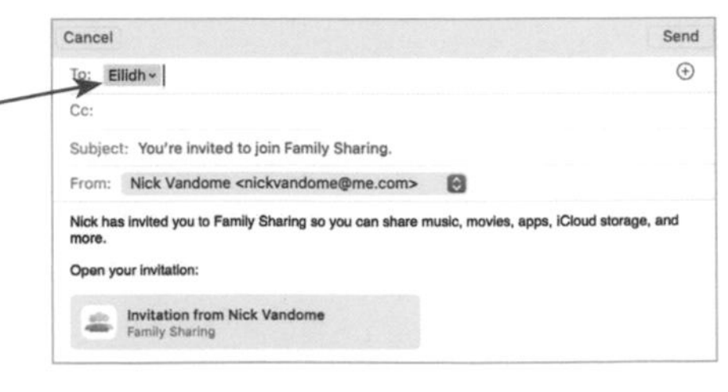

A Family Sharing invitation can be sent by email or using the Messages app, or by using AirDrop with another compatible Apple device in close proximity.

Using Family Sharing

Once you have set up Family Sharing and added family members, you can start sharing a selection of items.

Sharing photos

Photos can be shared using Family Sharing, with the Family album that is created automatically within the Photos app.

Click on the **Photos** app

Click on the **Shared Albums** button underneath the Albums heading in the left-hand sidebar

When a new item is added to a Family Sharing photo album, all of the Family Sharing members will be notified of this in their own Photos app.

The **Family** album is already available in the **Shared Albums** section. Double-click on the Family album to open it

Click on the **Add photos and videos** option

Add photos and videos

Click on the photos you want to add, and click on the **Add** button

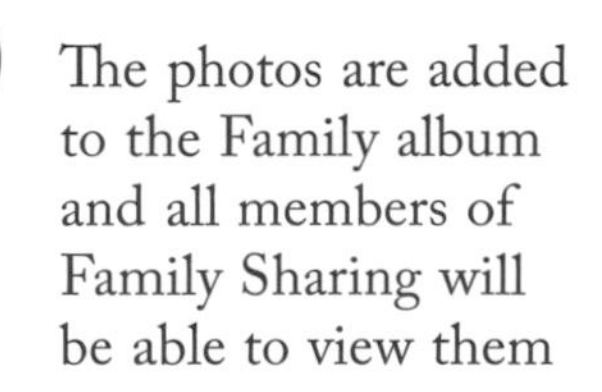

The photos are added to the Family album and all members of Family Sharing will be able to view them

...cont'd

Sharing apps, music, books and movies

Family Sharing means that all members of the group can share purchases from the iTunes Store, the App Store or Books. This is done from the Purchased section of each app.

More than one family member can use content in the Family Sharing group at the same time.

1. Open the relevant app and access the **Purchased** section. (For the App Store, click on the Account icon and click on the Purchased button; for the iTunes Store, click on the Purchased button on the bottom toolbar; for the Books app, click on the Account icon)

2. For all three apps, click on a member under **Family Purchases** to view their purchases and download them, if required, by tapping once on this button

Sharing calendars

Family Sharing also generates a Family calendar that can be used by all Family Sharing members.

When a Family Sharing calendar event is created, other people in your Family Sharing circle will have this event added to their Family calendar, and they will also be sent a notification.

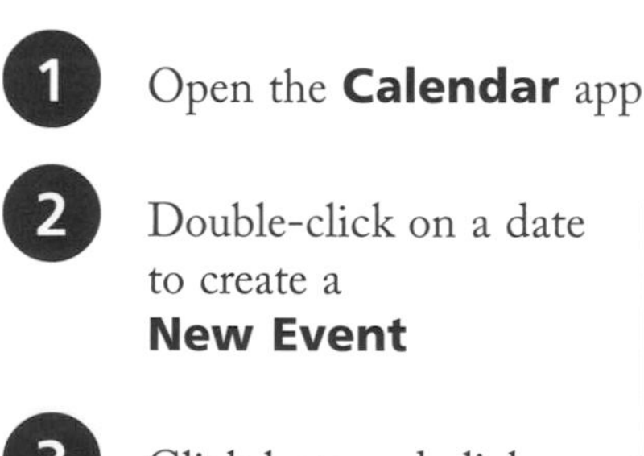

1. Open the **Calendar** app

2. Double-click on a date to create a **New Event**
3. Click here and click on the **Family** calendar
4. Complete the details for the event. It will be added to your calendar, with the Family color tag

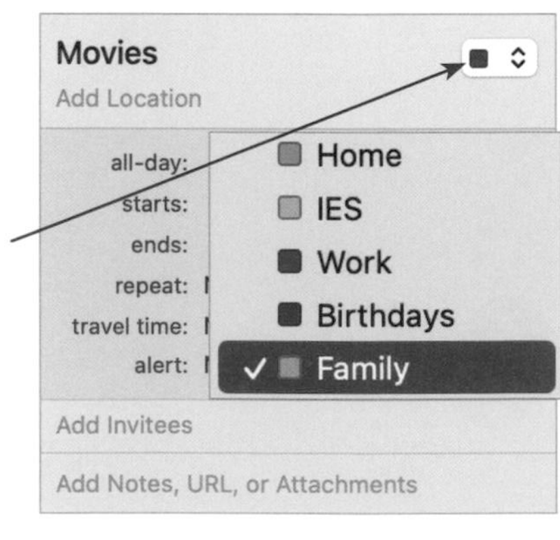

For more information about adding calendar events, see page 105.

Finding lost family devices

Family Sharing also makes it possible to see where everyone's devices are, which can be useful for locating people, but particularly so if a device belonging to a Family Sharing member is lost or stolen, including your own. To use this feature:

Ensure that the **Find My Mac** function is turned **On** in the **iCloud** system preferences, and log in to your online iCloud account at **www.icloud.com**

Click on the **Find iPhone** button (this works for other Apple devices, too)

Devices that are turned on, online and with iCloud activated are shown by green dots

4. Click on a green dot to display information about the device. Click on the **i** symbol to access options for managing the device remotely

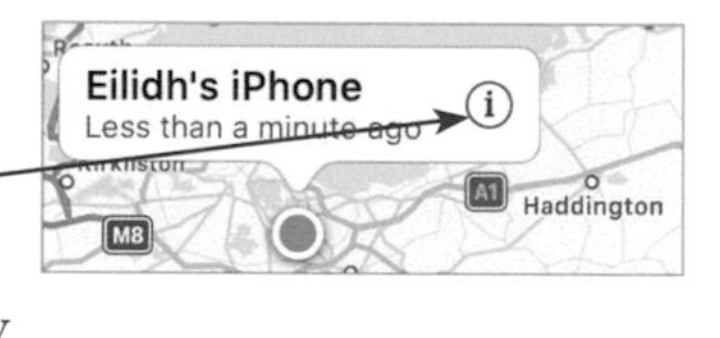

5. There are options to send an alert sound to the device, lock it remotely or erase its contents (if you are concerned about it having fallen into the wrong hands)

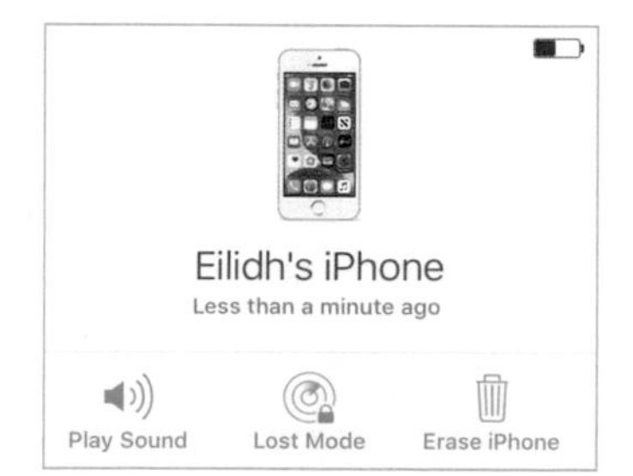

Other members of the Family Sharing group can add items to the Family calendar, and when they do, you will be sent a notification that appears on your Mac.

To change the color of a calendar, click on the **Calendars** button at the top left of the window. **Ctrl** + **click** on a calendar name and select a color from the bottom of the panel, or click on **Custom Color** to choose from the full spectrum.

Lost or missing Apple devices can be searched for from a Mac, using the **Find My** app. The missing devices have to be linked to the same Apple ID or be connected via Family Sharing.

Desktop Items

If required, the Desktop can be used to store apps and files. However, the Finder (see Chapter 3) does such a good job of organizing all of the elements within your computer that the Desktop is rendered largely redundant, unless you feel happier storing items here. The Desktop can also display any removable disks that are connected to your computer:

Icons for removable disks (e.g. flashdrives) will only become visible on the Desktop once a disk has been inserted.

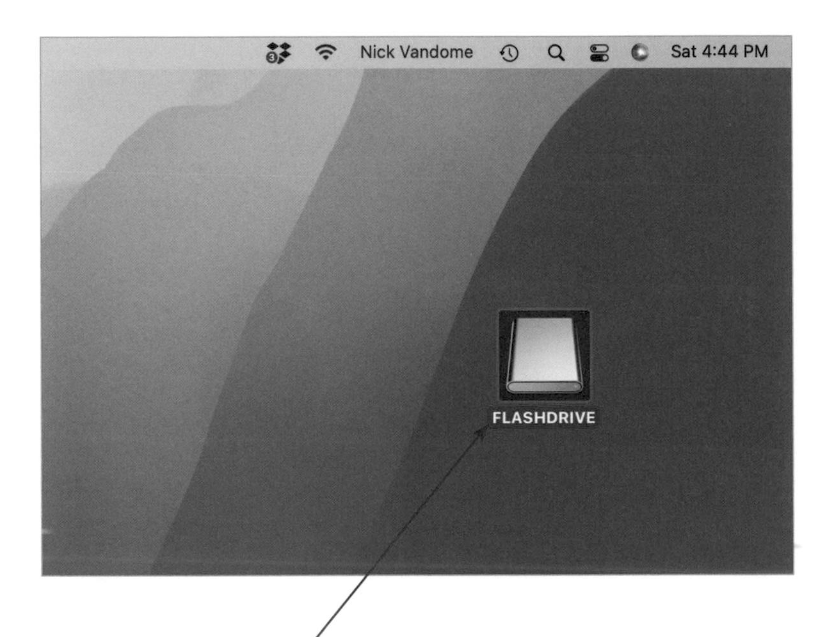

If a removable disk is connected to your computer, double-click the Desktop icon to view its contents.

Any removable disks that are connected to your computer can also be viewed by clicking on them in the sidebar in the Finder.

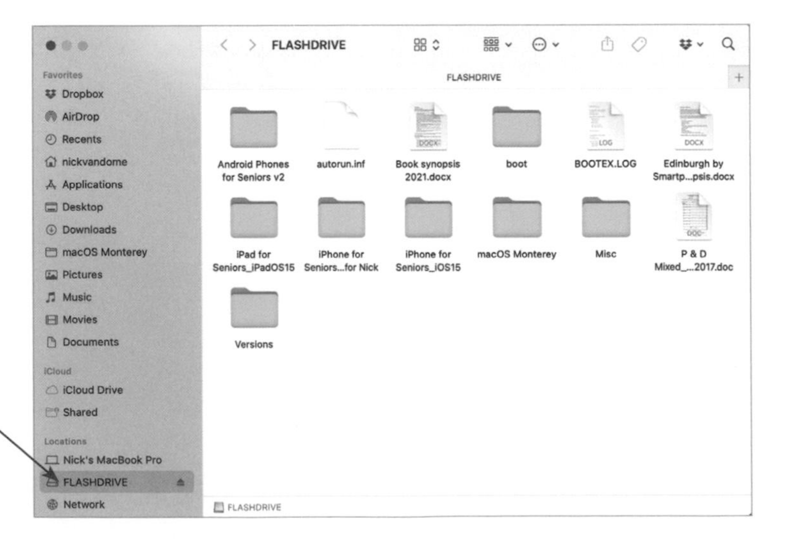

Ejecting Items

If you have removable disks attached to your Mac it is essential to be able to eject them quickly and easily. In macOS there are two ways in which this can be done:

1. In the Finder, click on this icon to the right of the name of the removable disk, or

2. On the Desktop, click and drag the disk icon over the **Trash**. This turns the Trash icon into the **Eject** icon, and the disk will be ejected

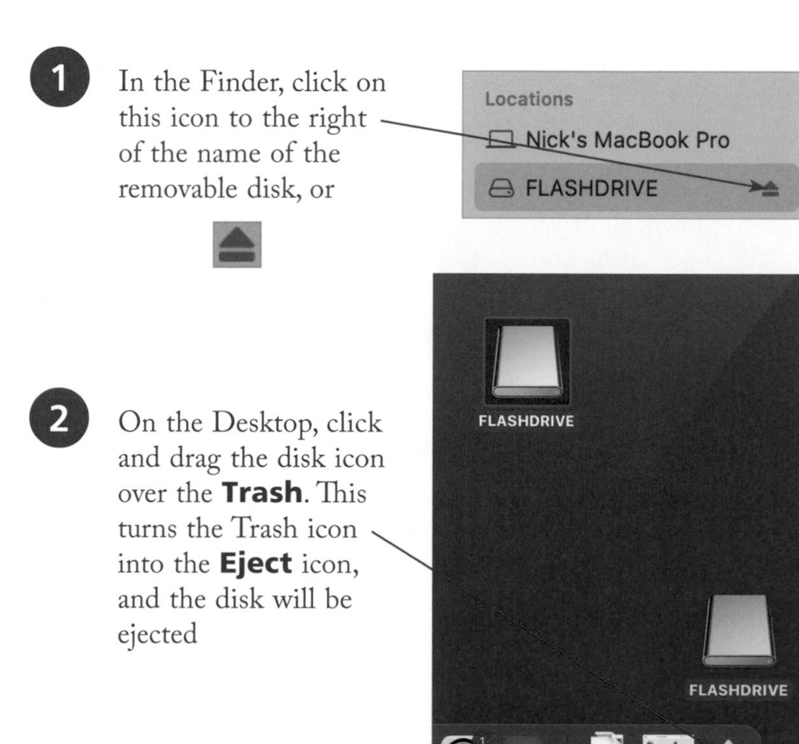

3. Some disks, such as CDs and DVDs (if they are connected with an external SuperDrive), are physically ejected when either of the above two actions is performed. Other disks, such as flashdrives, have to be removed manually once they have been ejected by macOS. If the disk is not ejected first, the following warning message will appear:

Disk Not Ejected Properly
Eject "FLASHDRIVE" before disconnecting or turning it off.

Hot tip

If the Desktop items are not showing, open the Finder and click on **Finder** > **Preferences** on the Menu bar and click on the **General** tab. Under **Show these items on the desktop** check **On** the items you want, including **Hard Disks**, **External Disks** and **CDs**, **DVDs**, **and iPods**.

Resuming

One of the chores about computing is that when you shut down your computer, you have to first close all of your open documents and apps and then open them all again when you turn your machine back on. However, macOS Monterey has a Resume feature that allows you to continue working exactly where you left off, even if you turn off your computer. To use this:

Before you shut down, all of your open documents and apps are available (as shown)

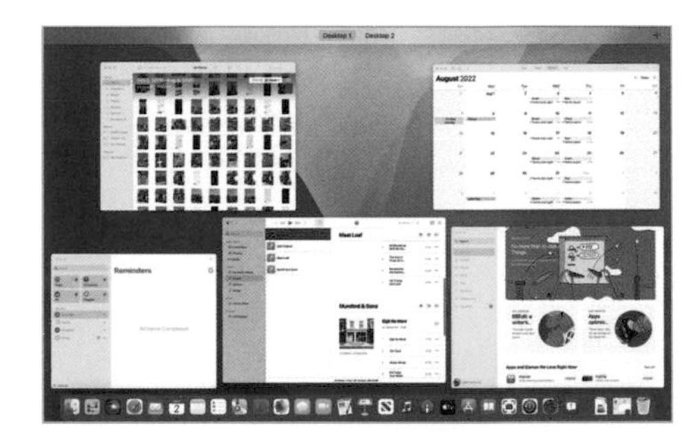

If you have open, unsaved documents you will be prompted to save them before your Mac is shut down.

Select the **Shut Down...** or **Restart...** option from the **Apple menu**

Finder File Edit View Go
About This Mac
System Preferences... 1 update
App Store... 11 updates
Recent Items
Force Quit...
Sleep
Restart...
Shut Down...

Make sure this box is checked **On** (this will ensure that all of your items will appear as before, after the Mac is shut down and then started up again)

Reopen windows when logging back in

Confirm the **Shut Down** or **Restart** command

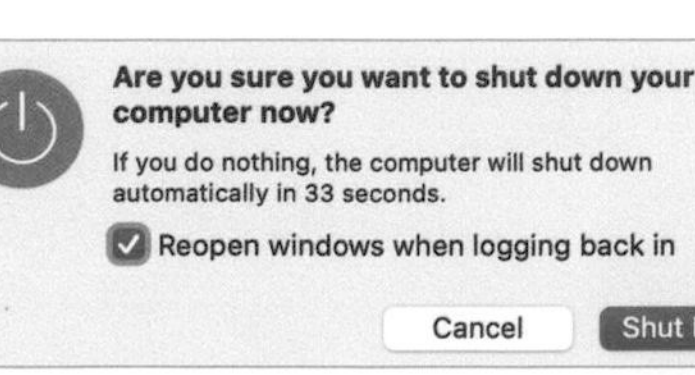

3 Finder

The principal method for moving around macOS Monterey is the Finder. This enables you to access items and organize your apps, folders and files. This chapter looks at how to use the Finder and how to get the most out of this powerful tool for navigating around macOS. It covers accessing items, how to customize the interface, and options for working with folders.

Working with the Finder

If you were only able to use one item on the Dock it would be the Finder. This is the gateway to all of the elements of your Mac. While it is possible to get to selected items through other routes, the Finder is the only location where you can gain access to everything on your system. If you ever feel that you are getting lost within macOS, click on the Finder and then you should begin to feel more at home. To access the Finder:

Don't forget

A link to iCloud Drive is also included in the Finder sidebar.

Click once on this icon at the left-hand side of the Dock

Overview

The Finder has its own toolbar; a sidebar from which items can be accessed; and a main window, where the contents of selected items can be viewed:

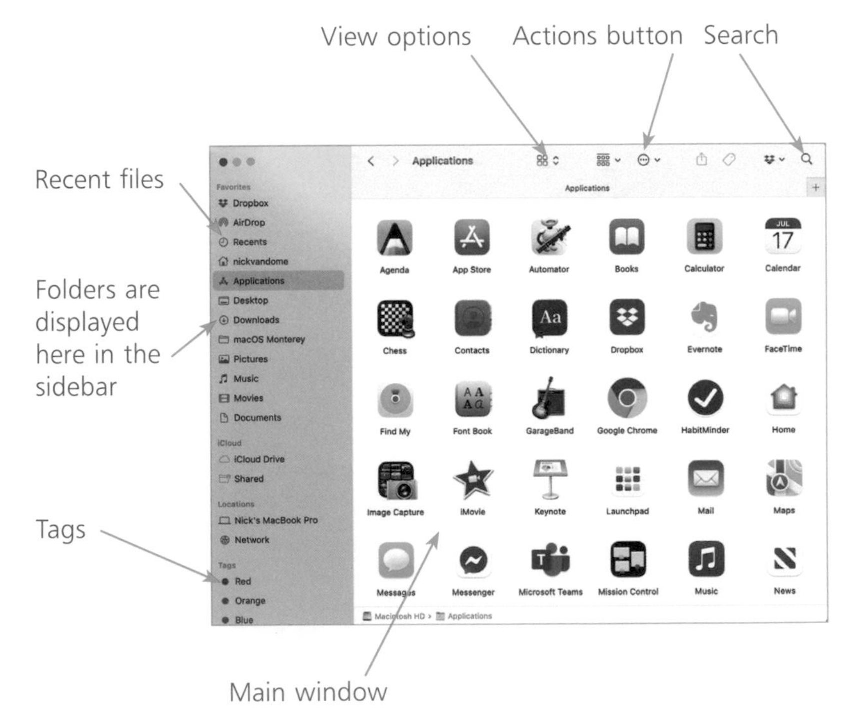

The **Actions** button has options for displaying information about a selected item and also options for how it is displayed within the Finder (see page 72 for details).

Tags in the Finder can be used to identify similar files and documents. To add tags, select the required items (see page 67) and drag a tag over them. To give tags specific names, **Ctrl + click** on one in the Finder sidebar, and click on the **Rename** option.

Finder Folders

Recents

This contains all of the latest files in which you have been working. They are sorted into categories according to file type so that you can search through them quickly. This is an excellent way to locate items without having to look through a lot of folders. To access this:

1. Click on this button in the Finder sidebar to access the contents of your **Recents** folder

2. All of your files are displayed in individual categories. Click on the headings at the top of each category to sort items by those criteria (in List view; see pages 59-62 for more details about Finder views)

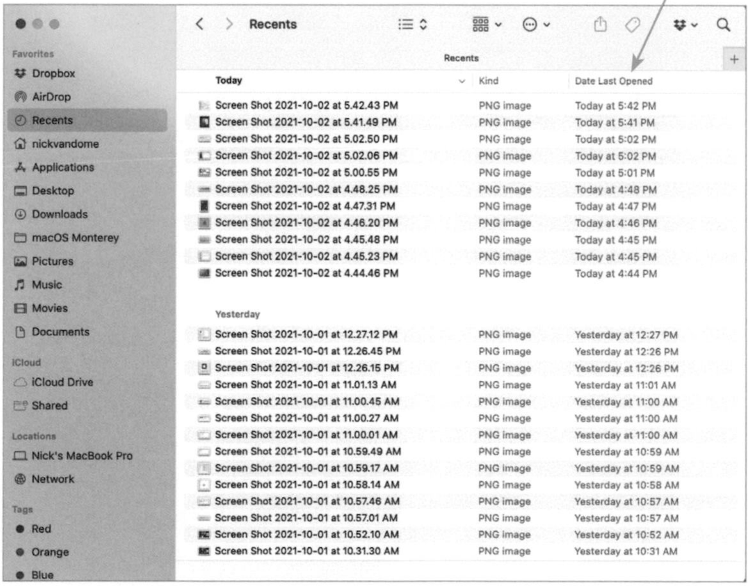

3. Double-click on an item to open it from the Finder

Don't forget

The Finder is always open (as denoted by the black dot underneath its icon on the Dock) and it cannot readily be closed down or removed.

Don't forget

The Finder sidebar has the macOS Monterey transparency feature so that you can see some of the open window, or Desktop, behind it.

Hot tip

To change the display of folders in the sidebar, click on the **Finder** menu on the top toolbar. Select **Preferences** and click on the **Sidebar** tab. Under **Show these items in the sidebar:** select the items you want included.

...cont'd

Home folder

This contains the contents of your own Home directory, containing your personal folders and files. macOS inserts some pre-named folders that it thinks will be useful, but it is possible to rearrange or delete these as you please. It is also possible to add as many more folders as you want.

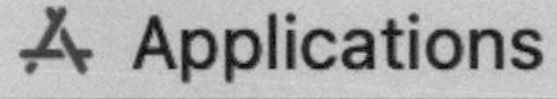

This folder contains all of the applications on your Mac. They can also be accessed from the Launchpad, as shown on pages 90-91.

Documents

This is part of your Home folder but is also on the Finder sidebar for ease of access. New folders can be created for different types of documents.

1. Click on your Apple ID account name in the sidebar to access the contents of your **Home** folder

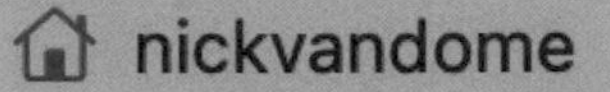

2. The Home folder contains the **Public** folder, which can be used to share files with other users if the computer is part of a network

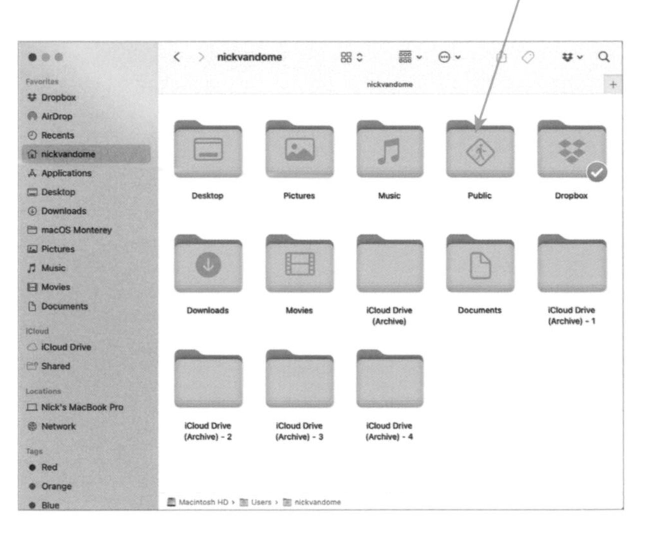

When you are creating documents, macOS by default recognizes their type and then, when you save them, suggests the most applicable folder in which to save them.

3. To add a new folder, click the **Actions** button (see page 72), then **New Folder** (or **Ctrl + click** in the window and select **New Folder**). The new folder will open in the main window, named "untitled folder". Overtype the name with one of your choice

4. To delete a folder from the Finder sidebar, **Ctrl + click** on it and select **Remove from Sidebar**

Finder Views

The way in which items are displayed within the Finder can be amended in a variety of ways, depending on how you want to view the contents of a folder. Different folders can have their own viewing options applied to them, and these will stay in place until a new option is specified.

Back button

When working within the Finder, each new window replaces the previous one, unless you open a new app. This prevents the screen becoming cluttered with dozens of open windows, as you look through various Finder windows for a particular item. To ensure that you never feel lost within the Finder structure, there is a Back button on the Finder toolbar that enables you to retrace the steps that you have taken.

Select an item within the Finder window and click on the space bar to view its details.

1. Navigate to a folder within the Finder (in this example, the **macOS Monterey** folder contained within **Pictures**)

2. Click on the **Back** button to move back to the previously-visited window (in this example, the main **Pictures** window)

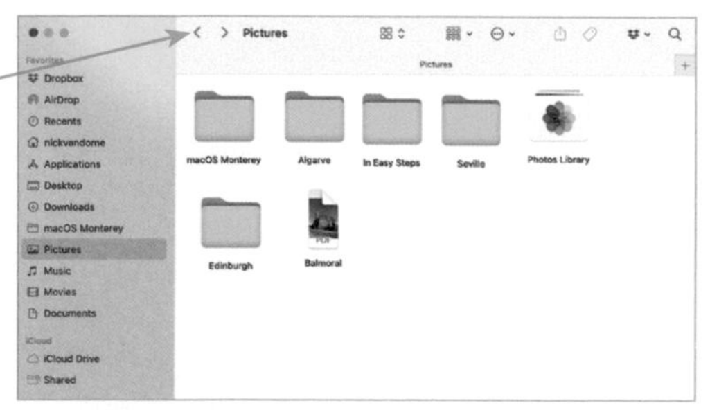

If you have not opened any Finder windows, the **Back** button will not operate.

...cont'd

View options

It is possible to customize some of the options for how items are viewed in the Finder. To do this:

1. Click on this button on the Finder toolbar

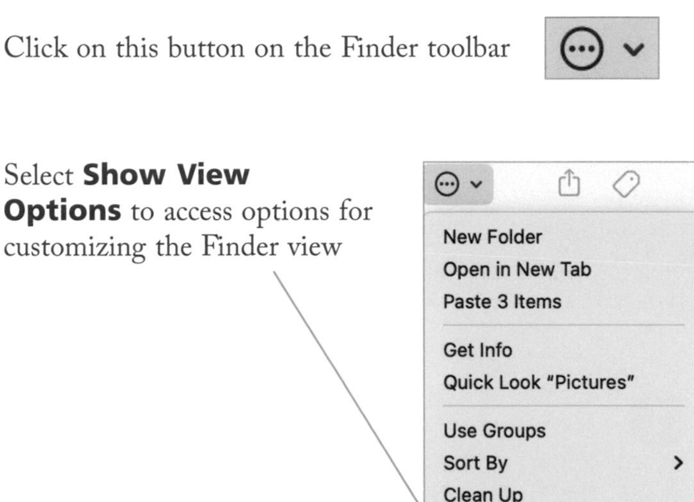

2. Select **Show View Options** to access options for customizing the Finder view

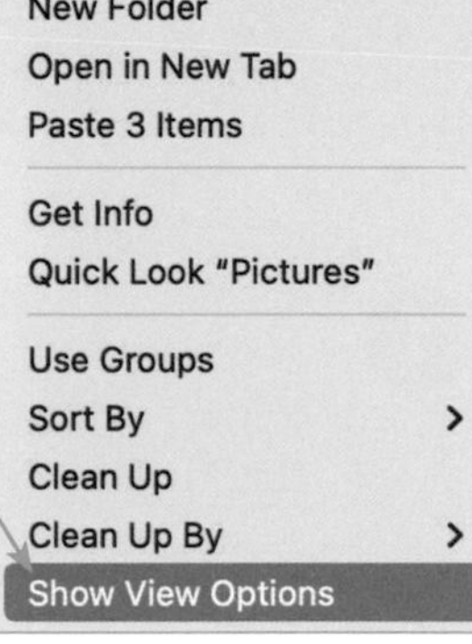

3. Drag this slider to set the icon size

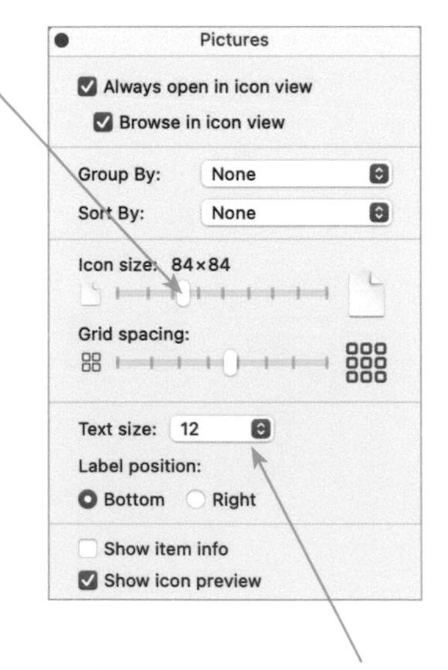

4. Select options here for the way items are arranged and displayed in Finder windows

A very large icon size can be useful for people with poor eyesight, but it takes up a lot more space in a window.

Icon, List and Columns views

There are several options for displaying folders and files within the Finder. To access them:

1. Click on this button on the Finder toolbar
2. Select one of the view options, from **as Icons**, **as List**, **as Columns** and **as Gallery** (see page 62 for Gallery view)

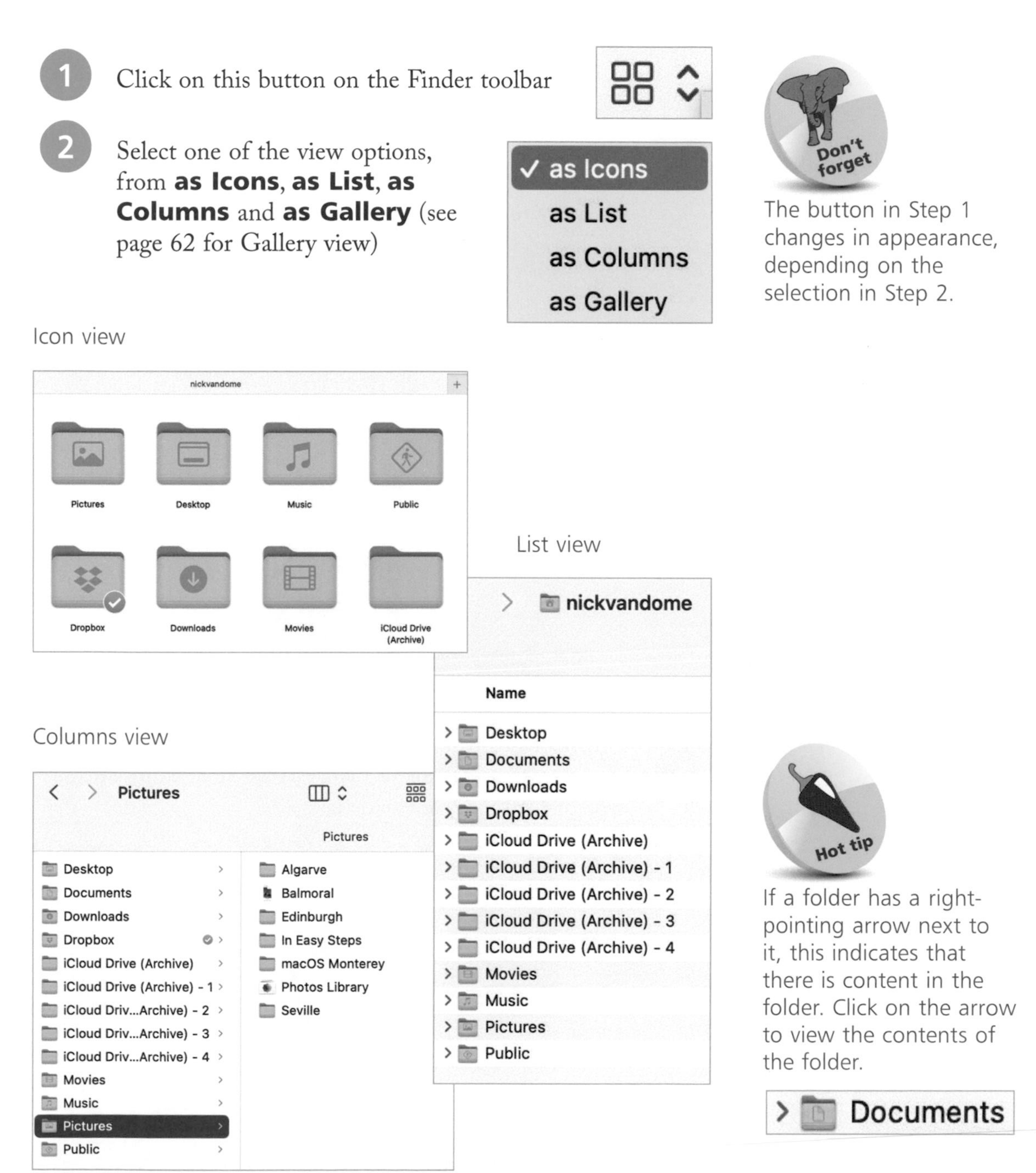

Icon view

List view

Columns view

Don't forget

The button in Step 1 changes in appearance, depending on the selection in Step 2.

Hot tip

If a folder has a right-pointing arrow next to it, this indicates that there is content in the folder. Click on the arrow to view the contents of the folder.

Gallery View

Another of the view options on the Finder toolbar is Gallery view: this is a view that has considerable functionality in terms of viewing details of specific files and also applying some quick editing functions to them without having to open the files. To use Gallery view:

1. Select a file in either Icon, List or Column view and click on the **as Gallery** option from the button on the top of the toolbar, as shown on page 61

✓ as Icons
as List
as Columns
as Gallery

2. The item is displayed in the main window, with thumbnails of the other available files below it

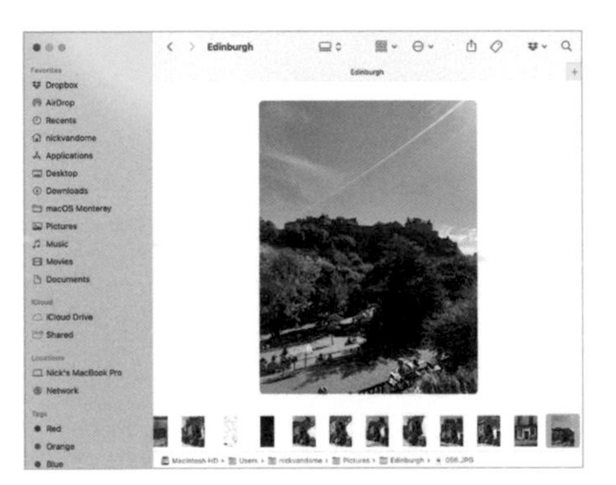

3. The item that is displayed will change depending on the type of file that is selected

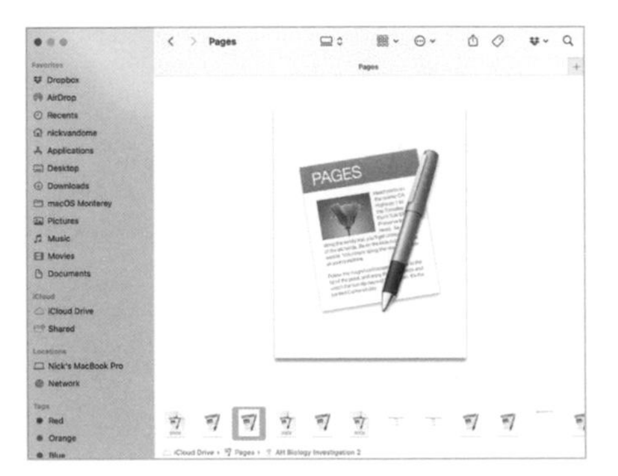

Gallery view is a good option for looking through photos without having to open the Photos app. Move through the photos by clicking on the thumbnails at the bottom of the screen, or with the arrow keys on the keyboard.

Quick Look

Through a Finder option called Quick Look, it is possible to view the content of a file without having to first open it. To do this:

1. Select a file in any of the Finder views

2. Press the space bar on the keyboard

8D6547A1-6132-4535-B72B-D5C8F..._1_105_c.jpeg

3. The contents of the file are displayed without it opening in its default program. Click on the **Markup** button on the top toolbar to add Markup items, such as drawing or writing

4. Click on the cross to close Quick Look, or view the item in full screen

The Markup button provides a range of options for drawing or writing on an item, including types of pens, shapes, borders, colors and text styles.

In Quick Look it is even possible to preview videos or presentations without having to first open them in their default app. If videos are being viewed, there will be a trim button so that the length of the video can be trimmed without having to open it in a specific app.

Finder Toolbar

Customizing the toolbar

As with most elements of macOS, it is possible to customize the Finder toolbar. To do this:

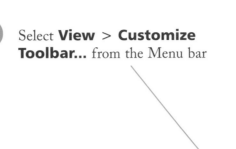

1. Select **View** > **Customize Toolbar...** from the Menu bar

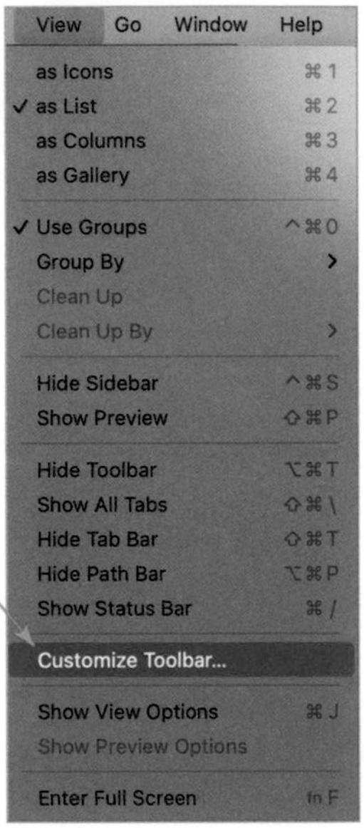

Do not put too many items on the Finder toolbar as you may not be able to see them all in the Finder window. If there are additional toolbar items, there will be a directional arrow indicating this. Click on the arrow to view the available items.

2. Drag individual items from the window into the toolbar

3. Alternatively, drag the default set of icons into the toolbar

4. Click **Done** at the bottom of the window

Finder Sidebar

Using the sidebar

The sidebar is the left-hand panel of the Finder that can be used to access items on your Mac. To use this:

1. Click on a button on the sidebar

2. Its contents are displayed in the main Finder window

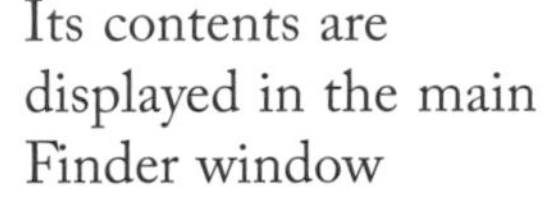

When you click on an item in the sidebar, its contents are shown in the main Finder window to the right.

Adding to the sidebar

Items that you access most frequently can be added to the sidebar.

1. Drag an item from the main Finder window onto the sidebar

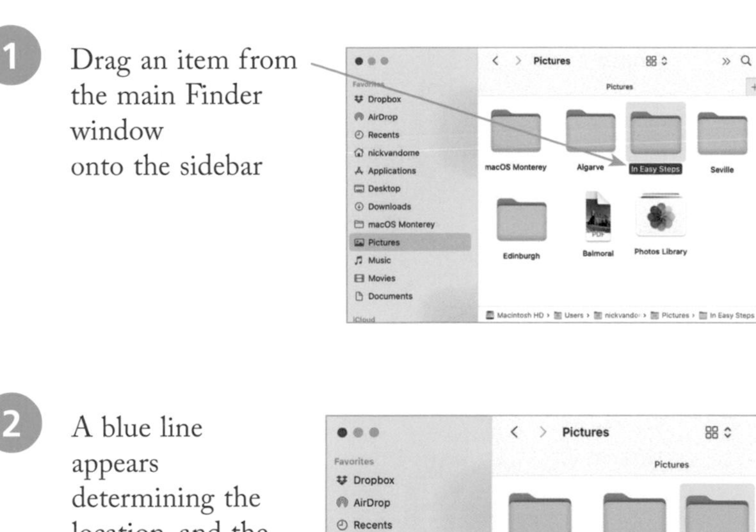

2. A blue line appears determining the location, and the item is added to the sidebar at this point. You can do this with apps, folders and files

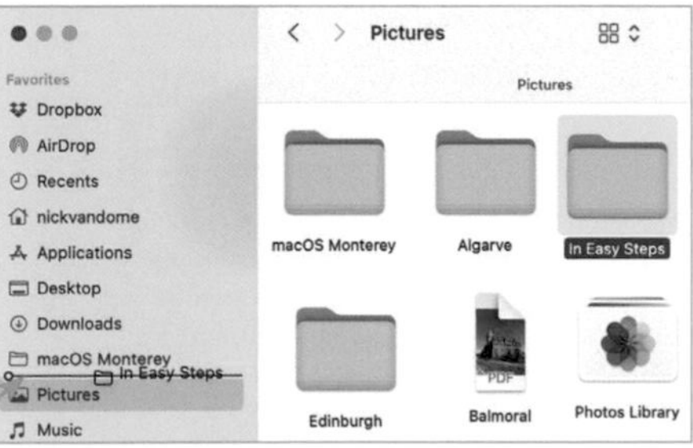

When items are added to the Finder sidebar, a shortcut – or alias – is inserted into the sidebar, not the actual item.

Don't forget

Items can be removed from the sidebar by **Ctrl + clicking** on them and selecting **Remove from Sidebar** from the contextual menu.

Working with Folders

Folders are always denoted by a folder icon. In Icon view they are larger than in List or Column views.

You can create as many "nested" folders (i.e. folders within other folders) as you want, although this can make your folder structure more complicated. Folders can also be dragged over other folders, and if you pause before you release the first folder you will be able to view the content of the target folder, to make sure this is where you want to place the first folder. This is known as spring-loaded folders.

Content can be added to an empty folder by dragging it from another folder and dropping it into the new one.

When macOS Monterey is installed, there are various folders that have already been created to hold apps and files. Some of these are essential (i.e. those containing apps), while others are created as an aid for where you might want to store the files that you create (such as the Pictures and Movies folders). Once you start working with macOS Monterey, you will probably want to create your own folders in which to store and organize your documents. This can be done on the Desktop or within any level of your existing folder structure. To create a new folder:

1. Access the location in which you want to create a new folder (e.g. your Home folder) and select **File** > **New Folder** from the Menu bar

2. A new, empty folder is inserted at the selected location (named "untitled folder")

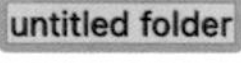

3. Overtype the file name with a new one. Press **Enter**

4. Double-click on the folder to view its contents (at this point it should be empty)

Selecting Items

Apps and files within macOS folders can be selected by a variety of different methods.

Selecting by dragging

Click and drag the cursor to encompass the items to be selected. The selected items will become highlighted.

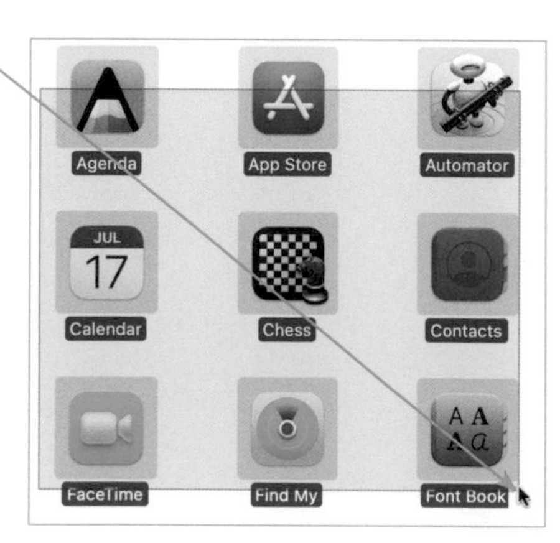

Don't forget

Once items have been selected, a single command can be applied to all of them. For instance, you can copy a group of items by selecting them and then applying the **Copy** command from the Menu bar.

Selecting by clicking

Click once on an item to select it, hold down Shift, and then click on another item in the list to select a consecutive group of items.

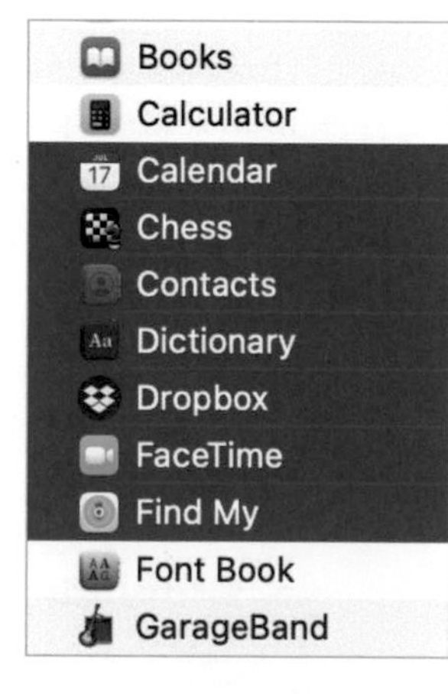

To select a non-consecutive group, select the first item by clicking on it once, then hold down the Command key (**cmd ⌘**) and select the other required items. The selected items will appear highlighted.

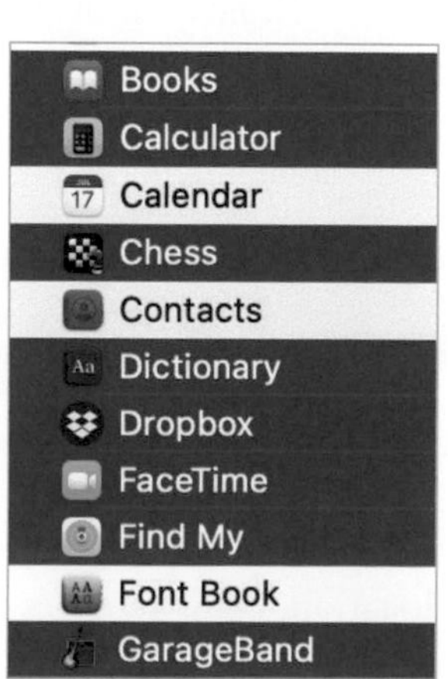

To select all of the items in a folder, select **Edit** > **Select All** from the Menu bar. The Select All command selects all of the elements within the active item. For instance, if the active item is a word processing document, the Select All command will select all of the items within the document; if it is a folder, it will select all of the items within that folder. You can also click **Command** + **A** to select all items.

Copying and Moving Items

Items can be copied and moved within macOS by using the copy and paste method, or by dragging.

When an item is copied, it is placed on the Clipboard and remains there until another item is copied.

Copy and paste

1. Select an item (or items) and select **Edit** > **Copy** from the Menu bar

Edit View Go Window
Undo ⌘Z
Redo ⇧⌘Z
Cut ⌘X
Copy 4 Items ⌘C
Paste ⌘V
Select All ⌘A

2. Move to the target location and select **Edit** > **Paste Item(s)** from the Menu bar. The item is then pasted into the new location

Edit View Go Window
Undo ⌘Z
Redo ⇧⌘Z
Cut ⌘X
Copy ⌘C
Paste 4 Items ⌘V
Select All ⌘A

3. To view items that have been copied onto the clipboard select **Edit** > **Show Clipboard** from the Menu bar. The items are displayed in the Clipboard window

Edit View Go Window
Undo ⌘Z
Redo ⇧⌘Z
Cut ⌘X
Copy ⌘C
Paste 4 Items ⌘V
Select All ⌘A
Show Clipboard

Clipboard
Screen Shot 2020-10-11 at 1.23.46 PM
Screen Shot 2020-10-11 at 1.24.32 PM
Screen Shot 2020-10-11 at 1.24.58 PM
Screen Shot 2020-10-11 at 1.25.41 PM

macOS Monterey supports the Universal Clipboard, where items can be copied on a device such as an iPhone and then pasted directly into a document on a Mac running macOS Monterey. The mobile device has to be running iOS 10 (or later) and the Mac has to support the Universal Clipboard. The process is the same as regular copy and paste, except that each operation is performed on each separate device and each app used has to be set up for iCloud. Also, Bluetooth and Wi-Fi have to be activated on both devices.

Dragging

Drag a file from one location to another to move it to that location. (This requires two or more Finder windows to be open, or drag the item over a folder on the sidebar.)

Finder Search

Searching electronic data is now a massive industry, with companies such as Google leading the way with online searching. On Macs it is also possible to search your folders and files, using the built-in search facilities: the Finder search; Siri (see pages 19-21); or Spotlight search (see pages 17-18).

Using Finder

To search for items within the Finder:

Try to make your search keywords and phrases as accurate as possible. This will create a better list of search results.

1. In the Finder window, click on this icon on the top toolbar and enter the search keyword(s) in the box. Search options are listed below the keyword

2. Select an option for which category you want to search over

3. The search results are shown in the Finder window. Click on one of these buttons to search specific areas

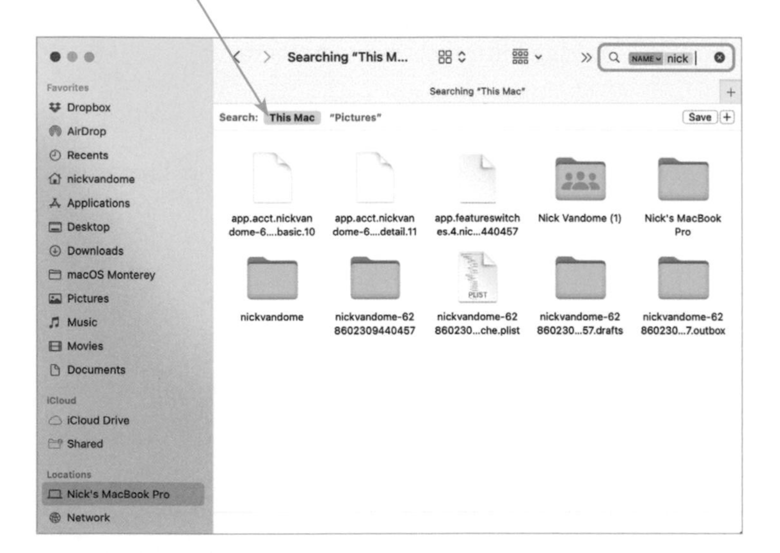

Both folders and files will be displayed in the Finder as part of the search results.

4. Double-click on an item to open it

Finder Tabs

Tabs in web browsers are now well established and allow you to have several pages open within the same browser window. This technology is included in the Finder in macOS Monterey, with Finder tabs. This enables different folders to be open in different tabs within the Finder so that you can organize your content exactly how you want. To do this:

1. Select **View** > **Show Tab Bar** from the Finder menu

2. A new tab appears at the right-hand side of the Finder

3. Click on this button to open the new tab

4. At this point, the content in the new tab is displayed in the window (see the second Hot tip)

Hot tip

In macOS Monterey, tabs are available in a range of apps that open multiple windows, including the Apple productivity apps: Pages, Numbers and Keynote. If the tabs are not showing, select **View** > **Show Tab Bar** from the app's Menu bar.

Hot tip

To specify an option for what appears in the default window for a new Finder window tab, click on the **Finder** menu and click on **Preferences**, then the **General** tab. Under **New Finder windows show:**, select the default window to be used.

5. Access a new folder to display this as the content for the tab. In this way, you can use different tabs for different types of content such as Recents, or photos, or for topics such as Work, Travel or Finance within Documents

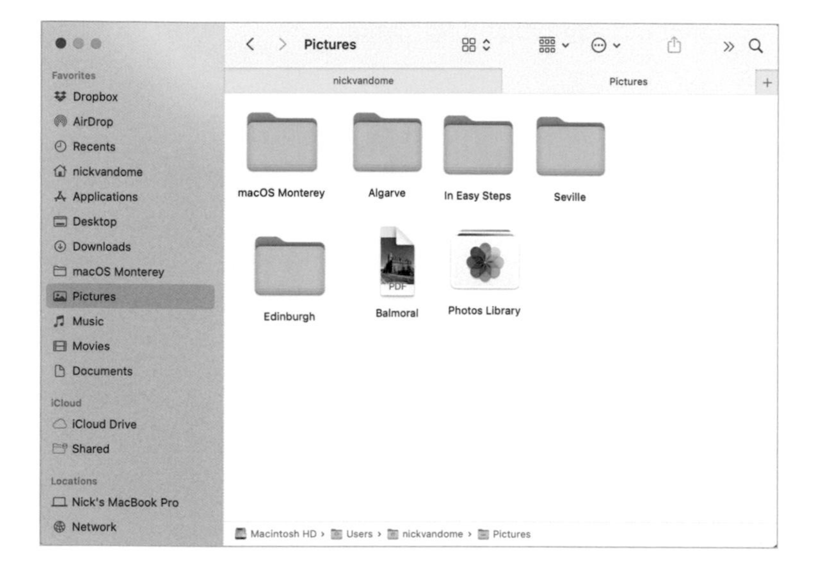

6. Each tab view can be customized, and this is independent of the other tabs

Dozens of tabs can be added in the Finder. However, when there are too many to fit along the Tab bar they are stacked on top of each other, so it can be hard to work out what you have in your tabs.

Actions Button

The Finder Actions button provides a variety of options for any item or items selected in the Finder. To use this:

The icons on the Finder toolbar can be changed by customizing them – see page 64.

Select an item or group of items about which you want to find out additional information

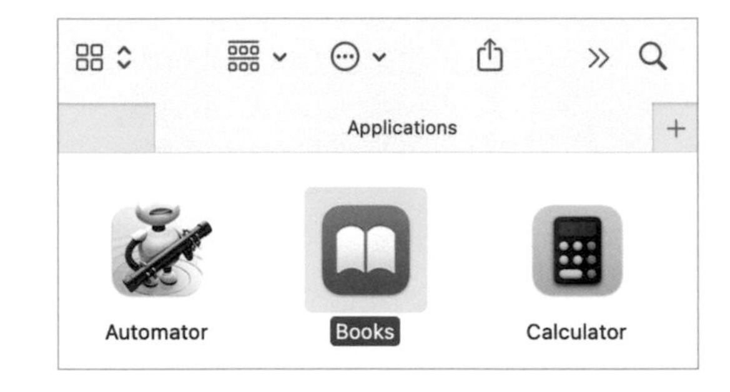

Click on the **Actions** button on the Finder toolbar

3 The available options for the selected item or items are displayed. These include **Get Info**, which displays additional information about an item such as file type, file size, creation and modification dates, and the default app for opening the item(s)

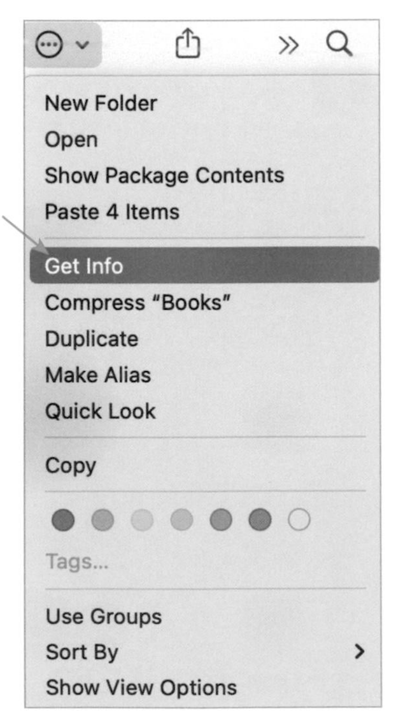

The **Actions** button can also be used for labeling items with Finder Tags. To do this, select the required items in the Finder and click one of the colored dots at the bottom of the **Actions** button menu. The selected tag will be applied to the item names in the Finder.

Sharing from the Finder

Also on the Finder toolbar is the Share button. This can be used to share a selected item or items in a variety of ways appropriate to the type of file that has been selected. For instance, a photo will have options including social media sites such as Twitter, while a text document will have fewer options. To share items directly from the Finder:

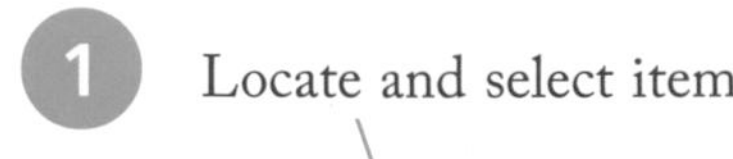

1. Locate and select item(s) that you want to share

2. Click on the **Share** button on the Finder toolbar

3. Select one of the options for sharing the item selected in Step 1

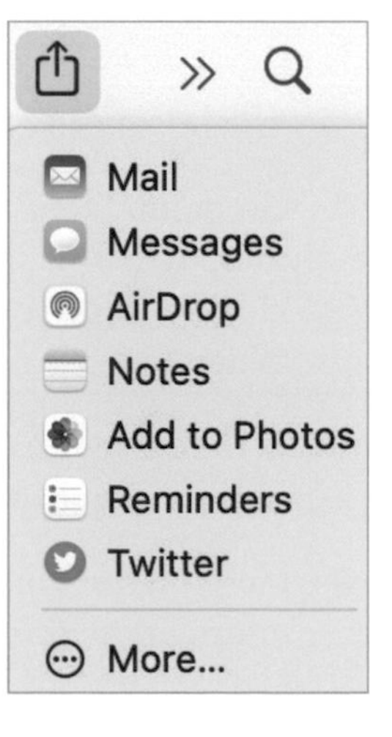

This button on the Finder is for changing the arrangement of items within the Finder. Click on this to access arrangement options such as Name, Date and Size:

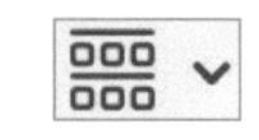

Menus

The main Apple Menu bar in macOS Monterey contains a variety of menus that are accessed when the Finder is the active window. When individual apps are open they have their own Menu bars, although in a lot of cases these are similar to the standard Menu bar, particularly for the built-in macOS Monterey apps such as Calendar, Contacts and Notes.

- **Apple menu**. This is denoted by a translucent blue apple and contains general information about the computer, links to System Preferences and the App Store for app updates, and options for shutting down your Mac.
- **Finder menu**. This contains preference options for amending the functionality and appearance of the Finder, and also options for emptying the Trash and accessing other apps (under the Services option).
- **File menu**. This contains common commands for working with open documents, such as opening and closing files, creating aliases, moving to the Trash, ejecting external devices, and burning discs.
- **Edit menu**. This contains common commands that apply to the majority of apps used on the Mac. These include Undo, Cut, Copy, Paste, Select All, and Show the contents of the Clipboard; i.e. items that have been cut or copied.
- **View menu**. This contains options for how windows and folders are displayed within the Finder, and for customizing the Finder toolbar. This includes showing or hiding the Finder sidebar and selecting view options for the size at which icons are displayed within Finder windows.
- **Go menu**. This can be used to navigate around your computer. This includes moving to your Recents folder, your Home folder, your Applications folder and recently-accessed folders.
- **Window menu**. This contains commands to organize currently-open apps and files on your Desktop.
- **Help menu**. This contains Mac Help files, which contain information about all aspects of macOS Monterey.

4 Navigating in macOS

macOS has Multi-Touch Gestures for navigating around your apps and documents. This chapter looks at how to use these to get around your Mac and also at using Mission Control.

The macOS Way of Navigating

One of the most revolutionary features of earlier versions of macOS, which is continued with macOS Monterey, is the way in which you can navigate around your applications, web pages and documents. This involves a much greater reliance on swiping on a trackpad or adapted mouse; techniques that have been imported from the iPhone and the iPad. These are known as Multi-Touch Gestures, and to take full advantage of these you will need to have one of the following devices:

- **A trackpad**. This can be found on MacBooks.
- **A Magic Trackpad**. This can be used with an iMac, a Mac mini or a Mac Pro. It works wirelessly via Bluetooth.
- **A Magic Mouse**. This can be used with an iMac, a Mac mini or a Mac Pro. It works wirelessly via Bluetooth.

All of these devices work using a swiping technique with fingers moving over their surface. This should be done with a light touch; it is a gentle swipe, rather than any pressure being applied to the device.

The trackpads and Magic Mouse do not have any buttons in the same way as traditional devices. Instead, specific areas are clickable so that you can still perform left- and right-click operations.

On a Magic Mouse, the center and right side can be used for clicking operations, and on a Magic Trackpad, the left and right corners can perform the same tasks.

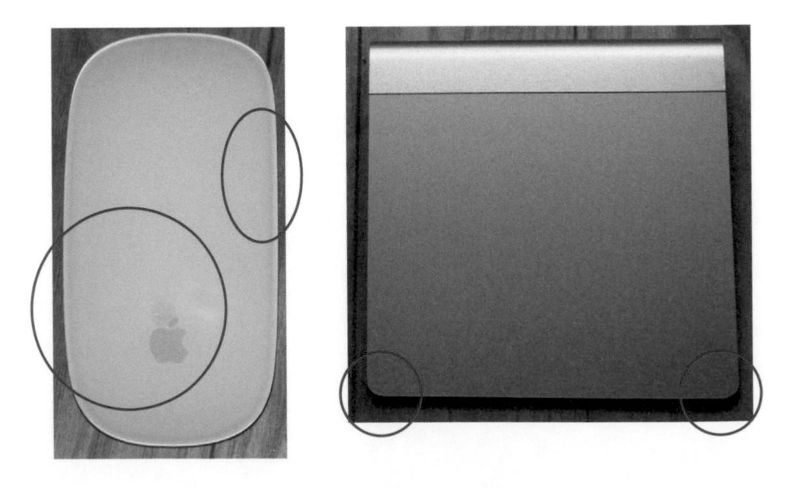

Hot tip

Models of MacBooks from May 2015 onward (and the Magic Trackpad) employ a technology known as Force Touch. This provides the user with different options, depending on how firmly they press on the trackpad. This means that more than one function can be performed, simply by pressing more firmly on the trackpad. It also provides haptic feedback, which is a physical response from the trackpad in the form of a small vibration, once different options have been accessed.

macOS Scroll Bars

In macOS Monterey, scroll bars in web pages and documents are more reactive to the navigational device being used on the computer. By default, with a Magic Trackpad, a trackpad or a Magic Mouse, scroll bars are only visible when scrolling is actually taking place. However, if a mouse is being used they will be visible permanently, although this can be changed for all devices.
To perform scrolling with macOS Monterey:

1. Scroll around a web page or document by swiping up or down on a Magic Mouse, a Magic Trackpad, or a trackpad. As you move up or down, the scroll bar appears

2. When you stop scrolling, the bar disappears to allow an optimum viewing area for your web page or document

3. To change the scroll bar options, select **System Preferences** > **General** and select the required settings under **Show scroll bars:**

Show scroll bars: ◉ Automatically based on mouse or trackpad
○ When scrolling
○ Always

If you do not have a trackpad, a Magic Trackpad or a Magic Mouse, you can still navigate in macOS Monterey with a traditional mouse and the use of scroll bars in windows.

For instructions on scrolling with a Magic Mouse, see page 84. For scrolling with a Magic Trackpad, see page 81.

Split View

When working with computers it can sometimes be beneficial to be able to view two windows next to each other. This can be to compare information in two different windows, or just to be able to use two windows without having to access them from the Desktop each time. In macOS Monterey, two windows can be displayed next to each other using the Split View feature.

1. By default, all open windows are layered on top of each other, with the active one at the top

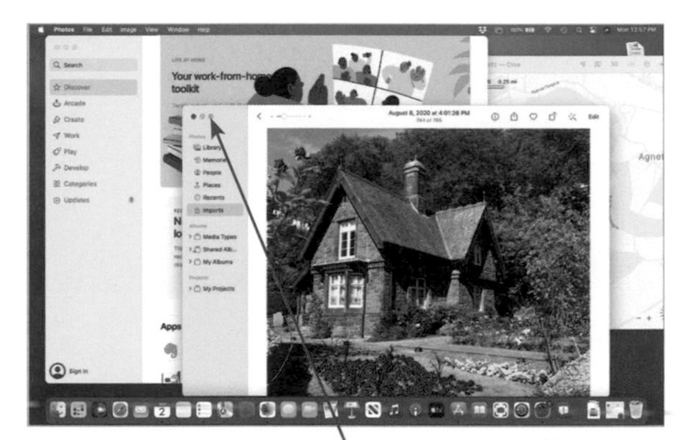

2. Press and hold on the green maximize button and select either **Tile Window to Left of Screen** or **Tile Window to Right of Screen** to position the window accordingly

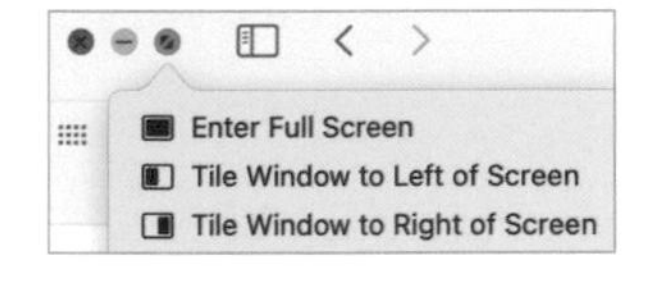

If you click once on the green maximize button this will display the app in full-screen mode, rather than holding on it to activate Split View.

Click on one of the apps on the right-hand side in Step 2 on the previous page to add it as the other Split View panel. Click on each window to make it active

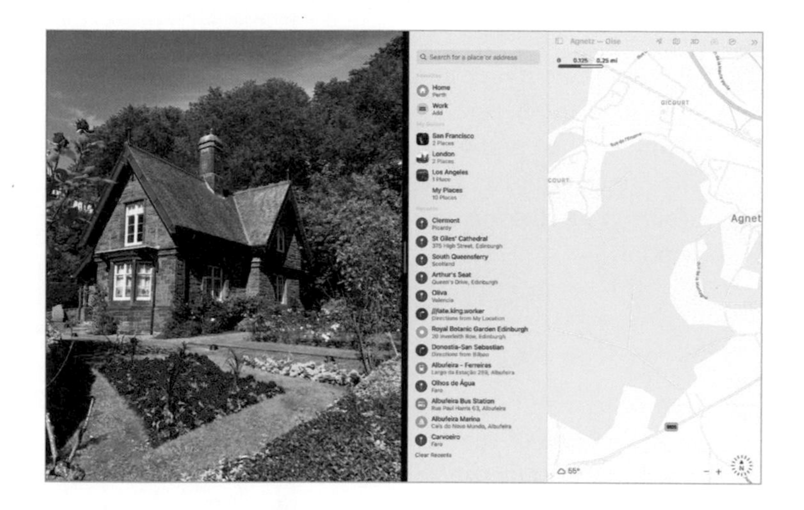

You can work in one panel in Split View (i.e. move through web pages) without affecting the content of the app on the other side.

Drag the middle divider bar to resize either of the Split View panels, to change the viewing area

Swap the windows in Split View by dragging the top toolbar of one app into the other window.

5 Move the cursor over the top of the window of each Split View item to view its toolbar and controls

Trackpad Gestures

Pointing and clicking

A Magic Trackpad, or trackpad, can be used to perform a variety of pointing and clicking tasks.

Tap with one finger in the middle of the Magic Trackpad or trackpad to perform a single-click operation; e.g. to click on a button or click on an open window

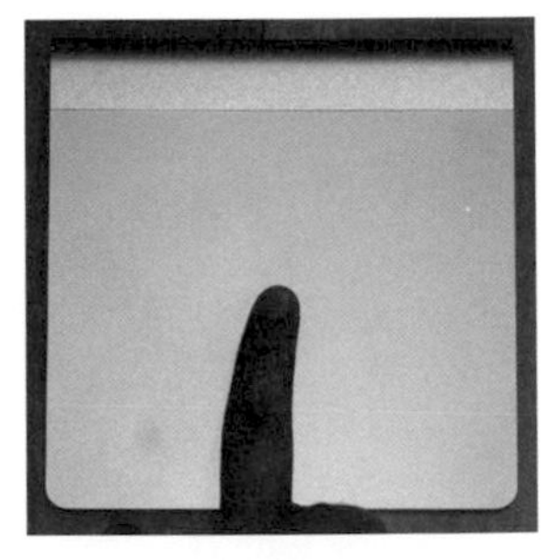

2 Tap once with two fingers in the middle of the Magic Trackpad or trackpad to access any contextual menus associated with an item (this is the equivalent of the traditional right-click with a mouse)

3 Press and hold with one finger to access data detectors for selected items. This is a content-specific function depending on the selected item: if it is a web page link it will show it in a pop-up window; if it is a calendar-related word – e.g. "tomorrow" – it will create a new calendar event

4 Swipe left with three fingers from the right-hand edge of the Magic Trackpad or trackpad to access the Notification Center

By default, the iMac with Retina Display comes with a Magic Mouse provided. However, this can be swapped for a Magic Trackpad, or both can be included for an additional cost.

Don't worry if you cannot immediately get the hang of Multi-Touch Gestures. It takes a bit of practice to get the correct touch and pressure on the Magic Trackpad, trackpad or Magic Mouse.

Scrolling and zooming

One of the most common operations on a computer is scrolling on a page, whether it is a web page or a document. Traditionally, this has been done with a mouse and a cursor. However, using a Magic Trackpad you can now do all of your scrolling with your fingers. There are a number of options for doing this.

Scrolling up and down

To move up and down web pages or documents, use two fingers on the Magic Trackpad or trackpad, and swipe up or down. The page moves in the opposite direction to the one in which you are swiping – i.e. if you swipe up, the page moves down, and vice versa.

1. Open a web page
2. Position two fingers in the middle of the Magic Trackpad or trackpad

3. Swipe them up to move down the page

4. Swipe them down to move up the page

If you have too many functions set using the same number of fingers, some of them may not work. See pages 85-86 for details about setting preferences for Multi-Touch Gestures.

When scrolling up and down pages, the gesture moves the page the opposite way – i.e. swipe down to move up the page, and vice versa.

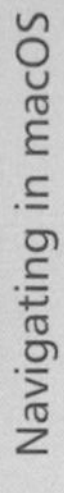

...cont'd

Zooming in and out

To zoom in or out on web pages or documents:

To zoom in, position your thumb and forefinger in the middle of the Magic Trackpad or trackpad

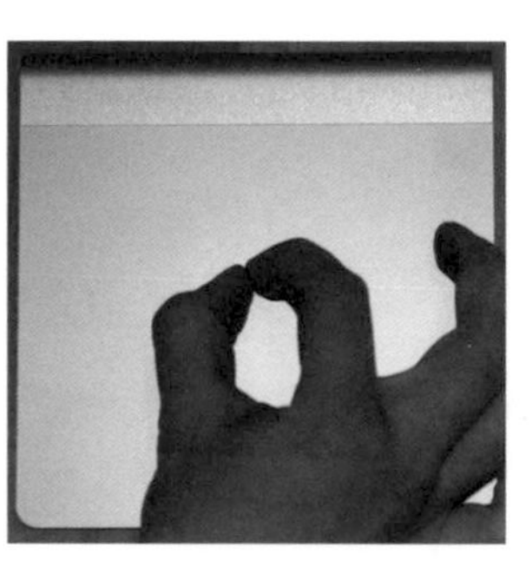

Don't forget

Pages can also be zoomed in on by double-tapping with two fingers.

Spread them outward to zoom in on a web page or document

There is a limit on how far you can zoom in or out on a web page or document, to ensure that it does not distort the content too much.

To zoom out, position your thumb and forefinger at opposite corners of the Magic Trackpad or trackpad

4 Pinch them into the center of the Magic Trackpad or trackpad, to zoom out

...cont'd

Moving between pages

With Multi-Touch Gestures, it is possible to swipe between pages within a document (such as a book). To do this:

Position two fingers to the left or right of the Magic Trackpad or trackpad

Swipe to the opposite side of the Magic Trackpad or trackpad, to move through the document

Moving between full-screen apps

In addition to moving between pages by swiping, it is also possible to move between different apps when they are in full-screen mode. To do this:

See pages 92-93 for details about using full-screen apps.

Position three fingers to the left or right of the Magic Trackpad or trackpad

Swipe to the opposite side of the Magic Trackpad or trackpad, to move through the available full-screen apps

Showing the Desktop

To show the whole Desktop, regardless of how many files or apps are open:

Position your thumb and three fingers in the middle of the Magic Trackpad or trackpad

2 Swipe to the opposite corners of the Magic Trackpad or trackpad, to display the Desktop

3 The Desktop is displayed, with all items minimized around the side of the screen

Magic Mouse Gestures

Pointing and clicking

A Magic Mouse can be used to perform a variety of pointing and clicking tasks.

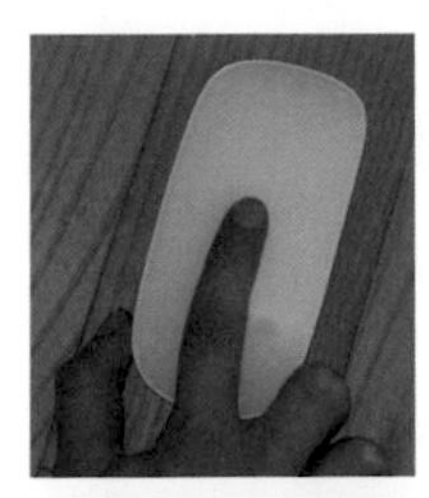

Click with one finger on the Magic Mouse to perform a single-click operation; e.g. to select a button or command

The right-click operation can be set within the Mouse system preferences.

Tap with one finger on the right side of the Magic Mouse to access any contextual menus associated with an item (this is the equivalent of the traditional right-click with a mouse)

Scrolling and zooming

The Magic Mouse can also be used to perform scrolling and zooming functions within a web page or document.

Swipe up or down with one finger to move up or down a web page or document. Double-tap with one finger to zoom in on a web page

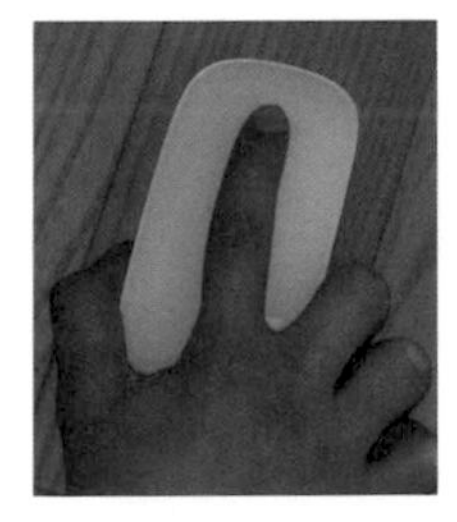

When scrolling on a web page or document, it moves in the opposite direction to the movement of your finger – i.e. if you swipe up, the page moves down, and vice versa.

Swipe left or right with one finger to move between pages

Swipe left or right with two fingers to move between full-screen apps

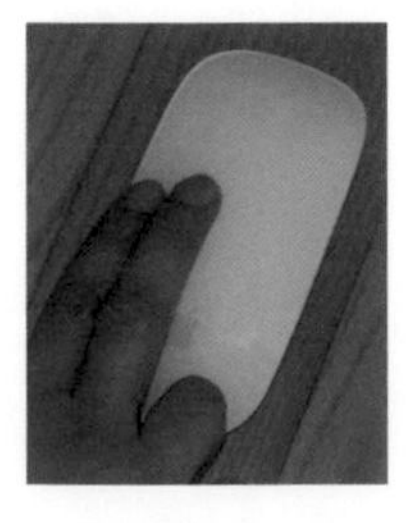

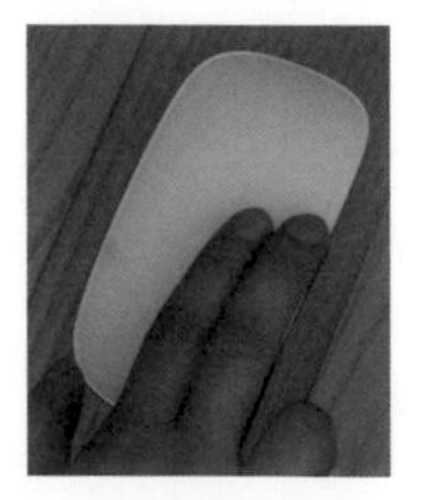

Multi-Touch Preferences

Some Multi-Touch Gestures only have a single action, which cannot be changed. However, others have options for changing the action linked to a specific gesture. This is done within the respective preferences for the Magic Mouse, the Magic Trackpad or the trackpad, where a full list of Multi-Touch Gestures is shown. To use these:

1. Access System Preferences and click on the Mouse or Trackpad button

2. Click on one of the tabs at the top

Point & Click | Scroll & Zoom | More Gestures

3. The actions are described on the left, with a graphic explanation on the right

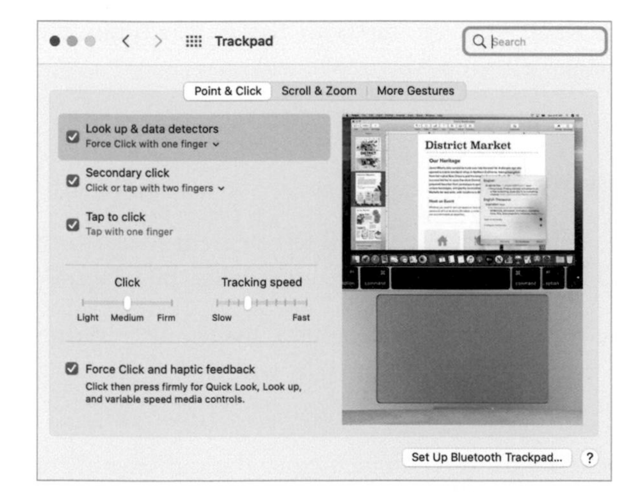

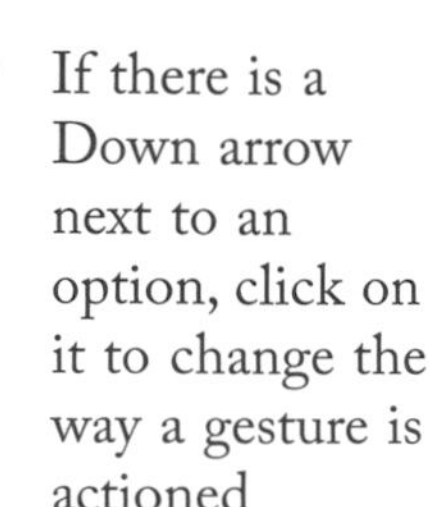

4. If there is a Down arrow next to an option, click on it to change the way a gesture is actioned

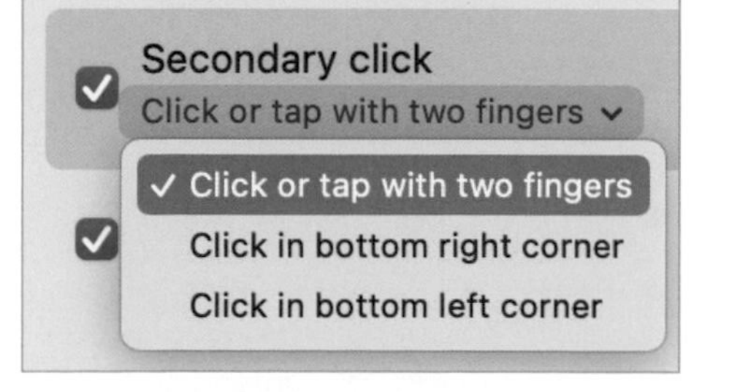

The Magic Trackpad, or trackpad, has three tabbed options within System Preferences: Point & Click, Scroll & Zoom, and More Gestures. The Magic Mouse has preferences for Point & Click and More Gestures.

...cont'd

Trackpad gestures

The full list of trackpad Multi-Touch Gestures, with their default actions are (relevant ones for the Magic Mouse are in brackets):

Point & Click

- Tap to click – tap once with one finger (same for the Magic Mouse).
- Secondary click – click or tap with two fingers (single-click on the right of the Magic Mouse).
- Lookup and data detectors – press and hold with one finger.

Scroll & Zoom

- Scroll direction: natural – with two fingers, content tracks finger movement (one finger with the Magic Mouse).
- Zoom in or out – spread or pinch with two fingers.
- Smart zoom – double-tap with two fingers (double-tap with one finger with the Magic Mouse).
- Rotate – rotate with two fingers.

More Gestures

- Swipe between pages – scroll left or right with two fingers (scroll left or right with one finger with the Magic Mouse).
- Swipe between full-screen apps – swipe left or right with three fingers (swipe left or right with two fingers with the Magic Mouse).
- Swipe left with three fingers from the right-hand edge of the trackpad or Magic Trackpad to access the Notification Center.
- Access Mission Control (see the next page) – swipe up with three fingers (double-tap with two fingers with the Magic Mouse).
- App Exposé – swipe down with three fingers.
- Access Launchpad – pinch with thumb and three fingers.
- Show Desktop – spread with thumb and three fingers.

Natural scrolling means the page follows the direction of your finger, vertically or horizontally, depending on which way you're scrolling.

Exposé enables you to view all of the active windows that are currently open for a particular app.

Mission Control

Mission Control is a function in macOS Monterey that helps you organize all of your open apps, full-screen apps, and documents. It also enables you to quickly view the Desktop. Within Mission Control there are also Spaces, where you can group similar types of documents together. To use Mission Control:

Click on a window in Mission Control to access it and exit the Mission Control window.

1. Click on this button in the Launchpad, or swipe upward with three fingers on the Magic Trackpad or trackpad, or double-tap with two fingers on a Magic Mouse

2. All open files and apps are visible via Mission Control

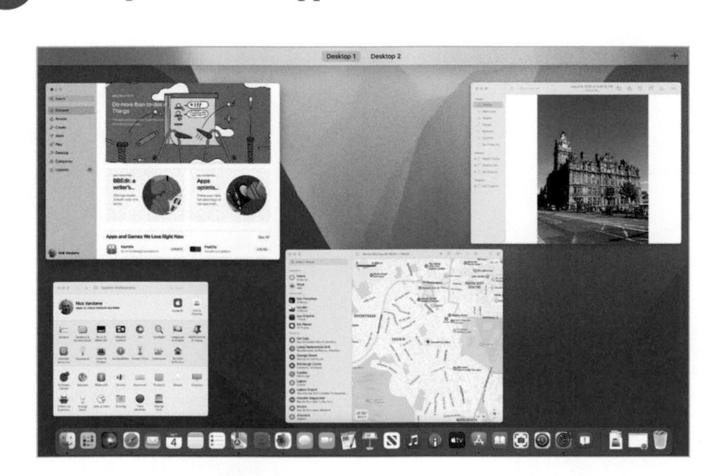

Any apps or files that have been minimized or closed do not appear within the main Mission Control window. Instead, they are located to the right of the dividing line on the Dock.

3. If there is more than one window open for an app they will each be shown separately

4. Move the cursor over the top of the Mission Control window to view the different Spaces (see page 88) and any apps in full-screen mode

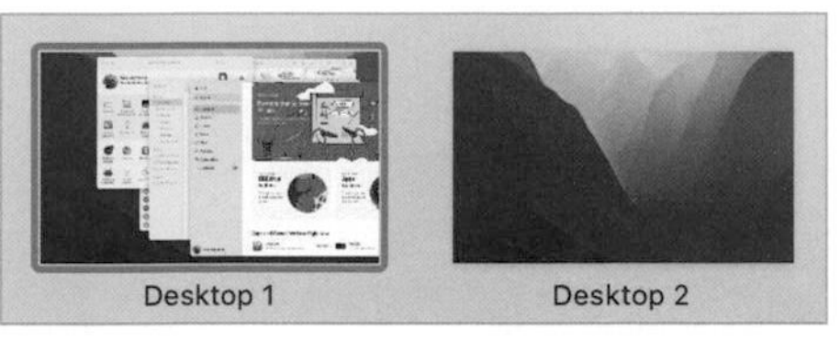

Drag an open app to the top of the screen to access Mission Control. You can then make an app full-screen from Mission Control by dragging it onto one of the Spaces at the top of the window.

Spaces

The top level of Mission Control contains Spaces, which are new Desktop areas into which you can group certain apps; e.g. the Apple productivity apps such as Pages and Numbers. This means that you can access these apps independently from other open items. This helps to organize your apps and files. To use Spaces:

Preferences for Spaces can be set within the Mission Control system preferences.

Move the cursor over the top right-hand corner of Mission Control and click on the **+** symbol

Create different Spaces for different types of content; e.g. one for productivity and one for entertainment.

A new **Space** is created along the top row of Mission Control

3. Drag an app onto the Space, and then any additional apps required

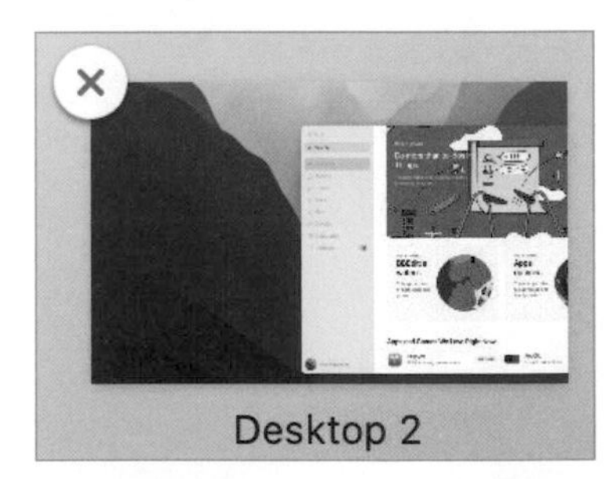

When you create a new Space it can subsequently be deleted by moving the cursor over it and clicking on the cross in the left-hand corner. Any items that have been added to this Space are returned to the default Desktop Space.

4. The new Space can then be accessed by clicking on the Space within Mission Control, and all of the apps that have been placed here, or opened within the Space, will be available

5 macOS Apps

Apps are the programs with which you start putting macOS to use, either for work or for fun. This chapter looks at using apps and the online App Store.

Launchpad

Even though the Dock can be used to store shortcuts to your applications, it is limited in terms of space. The full set of applications on your Mac can be found in the Finder (see Chapter 3) but macOS Monterey has a feature that allows you to quickly access and manage all of your applications. These include the ones that are pre-installed on your Mac and also any that you install yourself or download from the Apple App Store. This feature is called the Launchpad. To use it:

Hot tip

If the apps take up more than one screen, swipe from right to left with two fingers to view the additional pages, or click on the dots at the bottom of the window.

1 Click once on this button on the Dock

2 All of the apps (applications) are displayed

To launch an app from within the Launchpad, click on it once.

One of the apps in Utilities is Boot Camp Assistant, which can be used to run Windows on your Mac, if required.

3 Similar types of apps can be grouped together in individual folders. By default, the **Other** (**Utilities**) apps are grouped in this way

To create a group of similar apps, drag the icon for one over another

5. The apps are grouped together in a folder in the Launchpad, with a default name based on the type of app

To change the name, click on it once and overtype it with the new name

7. The folder appears within the **Launchpad** window

8. To remove an app, click and hold on it until it starts to jiggle and a cross appears. Click on the cross to remove it

Most system apps – i.e. the ones that already come with your Mac – cannot be removed in the Launchpad, only ones you have downloaded. The exceptions are Pages, Numbers, Keynote, iMovie and GarageBand.

Full-screen Apps

When working with apps we all like to be able to see as much of a window as possible. With macOS Monterey this is possible with the full-screen app option. This allows you to expand an app with this functionality so that it takes up the whole of your monitor or screen with a minimum of toolbars visible. Some apps have this functionality but some do not. To use full-screen apps:

By default, an app appears on the Desktop with other windows behind it

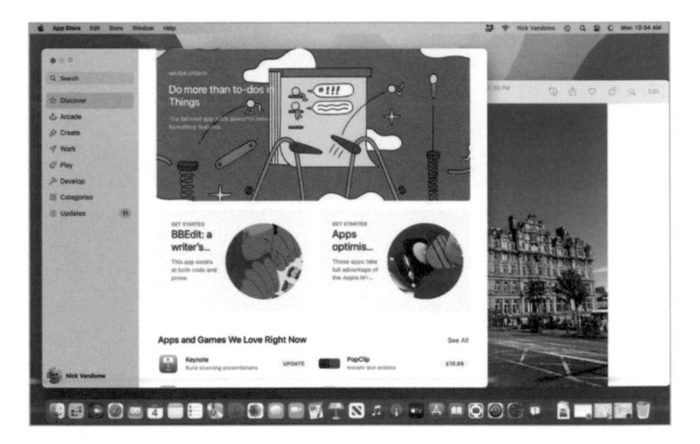

Click on this button in the top left-hand corner of the app's window

The app is expanded to take up the whole window. The main Apple Menu bar and the Dock are hidden

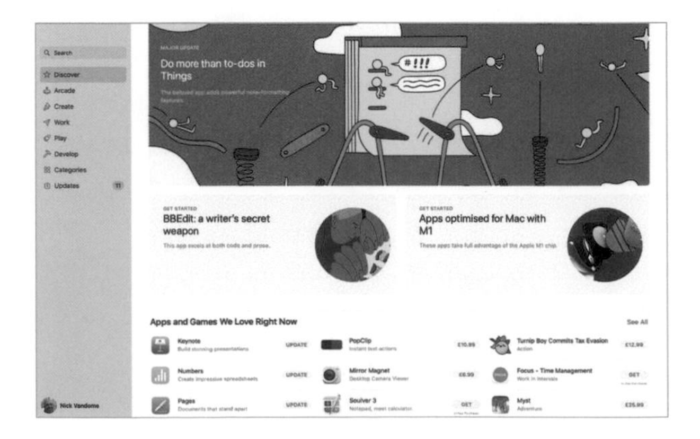

Don't forget

If the button in Step 2 is not visible then the app does not have the full-screen functionality.

4. To view the main Menu bar, move the cursor over the top of the screen

5. You can move between all full-screen apps by swiping with three fingers left or right on a trackpad or Magic Mouse

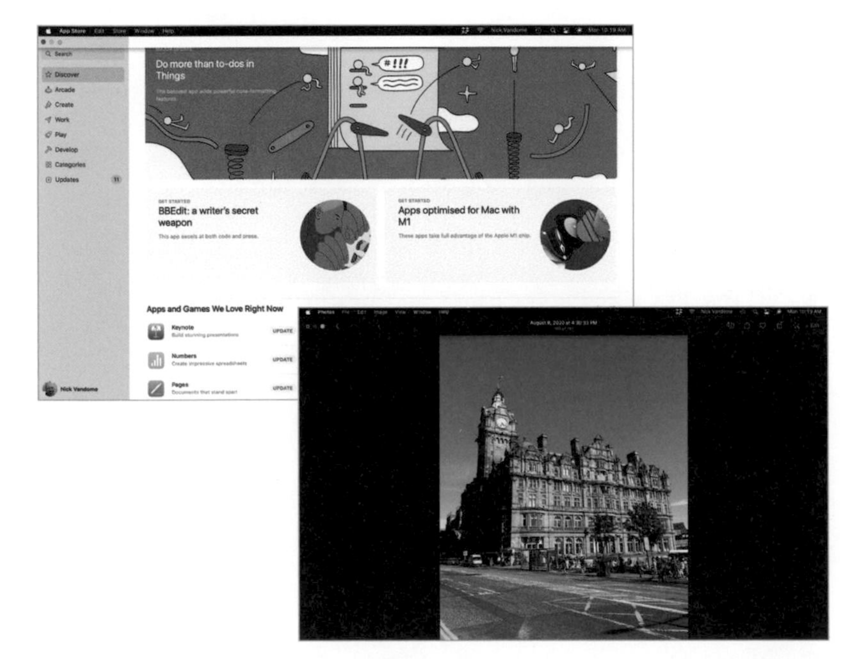

6. Move the cursor over the top left-hand corner of the screen and click on this button to close the full-screen functionality

7. In Mission Control all of the open full-screen apps are shown in the top row

For more information about navigating with Multi-Touch Gestures, see pages 80-86.

macOS Apps

macOS Monterey apps include:

- **Books**. An app for downloading ebooks. (See pages 155-156.)
- **Calculator**. A basic calculator.
- **Calendar**. (See pages 104-105.)
- **Contacts**. (See pages 102-103.)
- **Dictionary**. A digital dictionary.
- **FaceTime**. Can be used for video calls. (See pages 138-142.)
- **Find My**. This can be used to locate missing Apple devices.
- **Home**. This can be used to control compatible smart home devices; e.g. smart lighting and smart heating.
- **Mail**. The default email app.
- **Maps**. For viewing locations and destinations worldwide.
- **Messages**. (See pages 134-136.)
- **Mission Control**. The function for organizing your Desktop.
- **Music, TV, Photos and Podcasts**. (See Chapter 8 for details about these apps.)
- **News**. This aggregates news stories from several sources. (See page 112.)
- **Notes**. (See pages 106-109.)
- **Photo Booth**. An app for creating photo effects.
- **Preview**. (See page 118.)
- **Reminders**. (See pages 110-111.)
- **Safari**. The macOS-specific web browser.
- **Stocks**. This can be used to display stock market information.
- **Time Machine**. macOS's backup facility. (In some instances, this can be located in the Utilities/Other folder.)
- **Voice Memos**. (See page 113.)

Don't forget

The macOS apps can be accessed from the Launchpad and also the Applications folder from within the Finder.

Don't forget

Some apps are grouped together in the Utilities/Other folder. These include TextEdit, Chess, Stickies, Font Book, QuickTime Player, and Screenshot.

Macs with macOS Monterey also come with the Apple productivity apps, Pages (word processing), Numbers (spreadsheets) and Keynote (presentations). They are pre-installed and can also be downloaded, for free, from the App Store.

Accessing the App Store

The App Store is another macOS app. This is an online facility where you can download and buy new apps. These cover a range of categories such as productivity, business and entertainment. When you select or buy an app from the App Store, it is downloaded automatically by the Launchpad and appears there next to the rest of the apps.

To buy apps from the App Store, you need to have an Apple ID. If you have not already set this up, it can be done when you first access the App Store. To use the App Store:

You can set up an Apple ID when you first set up your Mac or you can do it when you first access the App Store.

1. Click on this icon on the Dock or within the Launchpad

2. The Homepage of the App Store contains the current top featured and best new apps

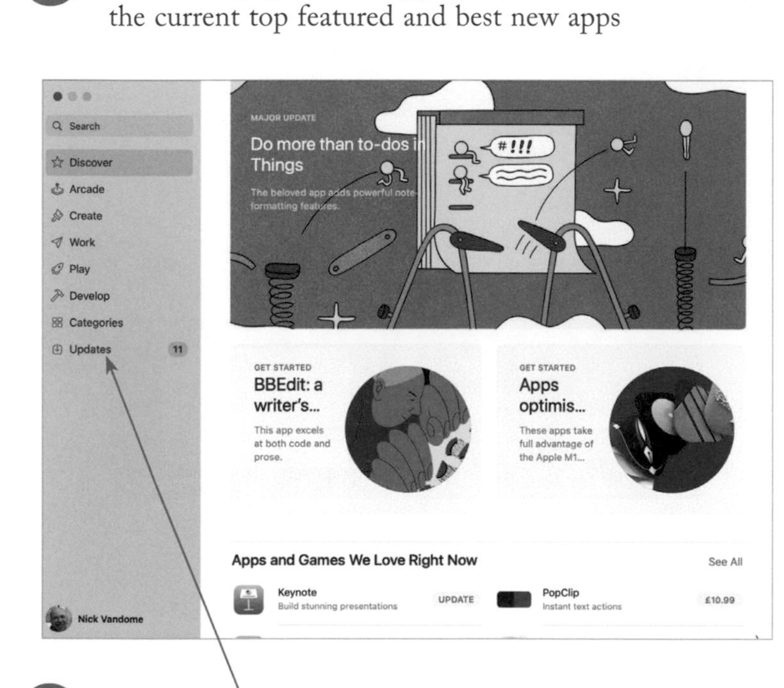

3. Your account information and quick links to other areas of the App Store are listed in the left-hand sidebar

Click on the **Arcade** button in the left-hand sidebar to access the Arcade option for playing games. This is a subscription service (with a one-month free trial) that gives unlimited access to original Apple games that can be played on your own, or with other people. It can also be used with Family Sharing.

By default, the App Store opens at the Homepage, which is accessed by clicking on the **Discover** button in the left-hand sidebar.

Downloading Apps

The App Store contains a wide range of apps: from small, fun apps to powerful productivity ones. However, downloading them from the App Store is the same regardless of the type of app. The only differences are whether they need to be paid for or not and the length of time they take to download. To download an app from the App Store:

1. Browse through the App Store until you find the required app

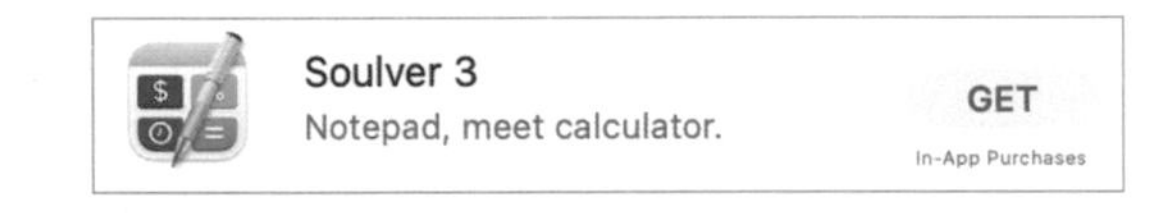

When downloading apps, start with a free one first so that you can get used to the process before you download paid-for apps.

2. Click on the app to view a detailed description about it

3. Click on the button next to the app icon to download it. If the app is free the button will say **Get**

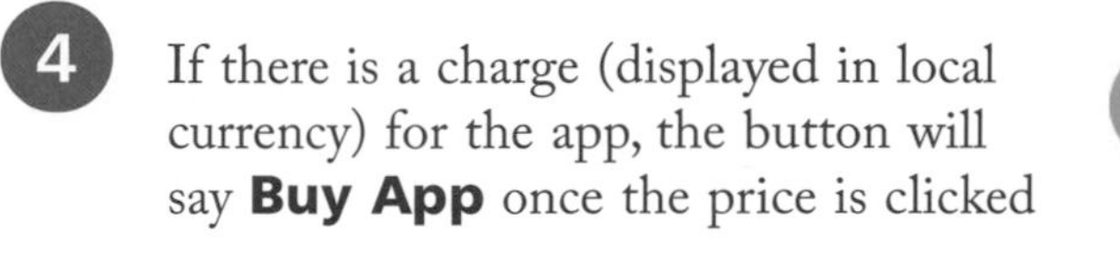

4. If there is a charge (displayed in local currency) for the app, the button will say **Buy App** once the price is clicked

5. Click on the **Install** button

6. Enter your **Apple ID** account details to continue downloading the app, then click on the **Get** button

Sign-In Required
If you have an Apple ID and password, enter them here. If you have used the iTunes Store or iCloud, for example, you have an Apple ID.
Apple ID: nickvandome@mac.com
Password: ••••••••
Forgot Apple ID or Password?
Cancel Get

Depending on their size, different apps take differing amounts of time to be downloaded.

7. The progress of the download is displayed in a progress bar underneath the Launchpad icon on the Dock

8. Once it has been downloaded, the app is available within the Launchpad

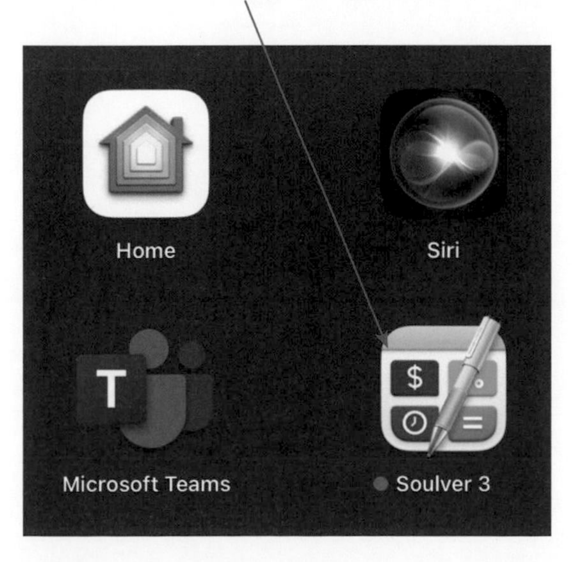

As you download more apps, additional pages will be created within the Launchpad to accommodate them.

Finding Apps

There are thousands of apps in the App Store and sometimes the hardest task is locating the ones you want. However, there are a number of ways in which finding apps is made as easy as possible.

Click on the **Discover** button

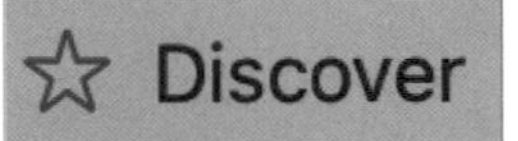

The main window has a range of suggested apps. Scroll down the page to see a panel with the current **Top Free** apps and games

Underneath this is a list of the current **Top Paid** apps and games

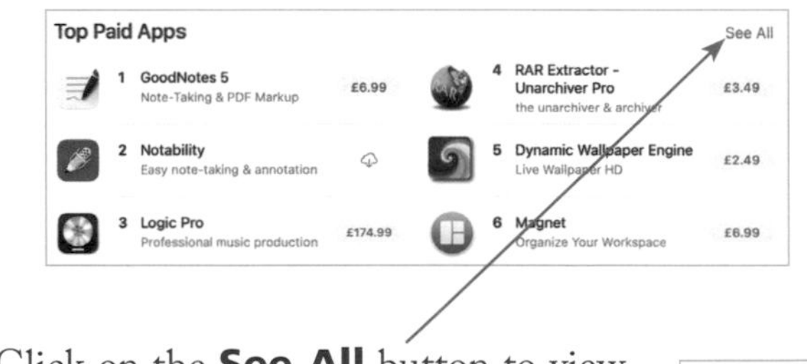

Click on the **See All** button to view the full list of Top Charts apps

See All

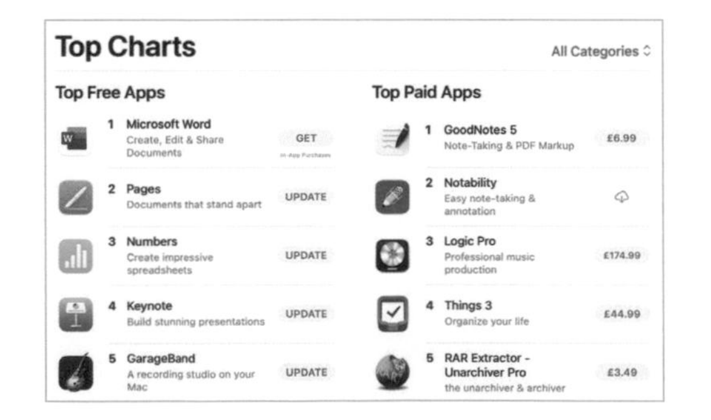

The Top Charts has sections for paid-for apps and free ones.

5. Click on the buttons in the left-hand panel to view the available apps with them

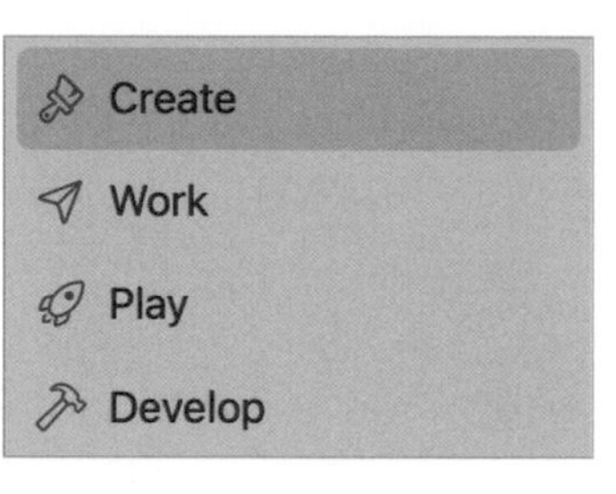

6. Each heading contains relevant apps

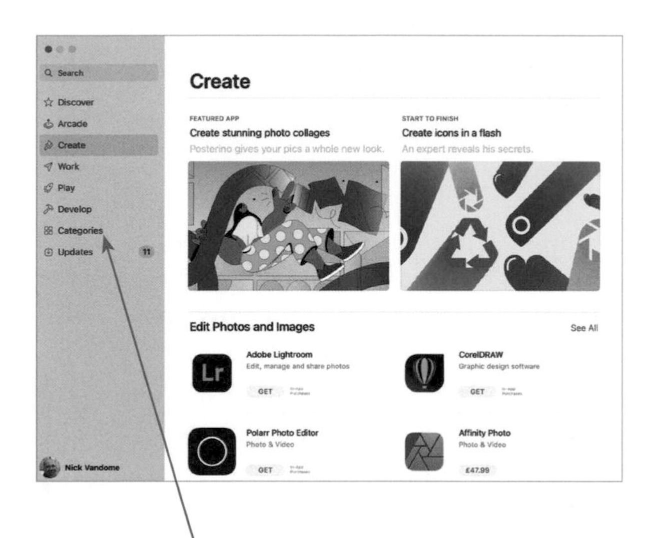

7. Click on the **Categories** button

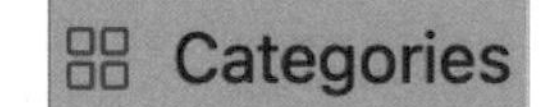

8. Browse through the apps by specific categories, such as Business, Entertainment and Finance

Another way to find apps is to type a keyword into the Search box in the top left-hand corner of the App Store window.

Managing Your Apps

Once you have bought apps from the App Store you can view details of ones you have purchased and also install updated versions of them. To view your purchased apps:

Even if you interrupt a download and turn off your Mac you will still be able to resume the download when you restart your computer.

1. Click on your own account name, in the bottom left-hand corner

2. Details of your purchased apps are displayed (including those that are free)

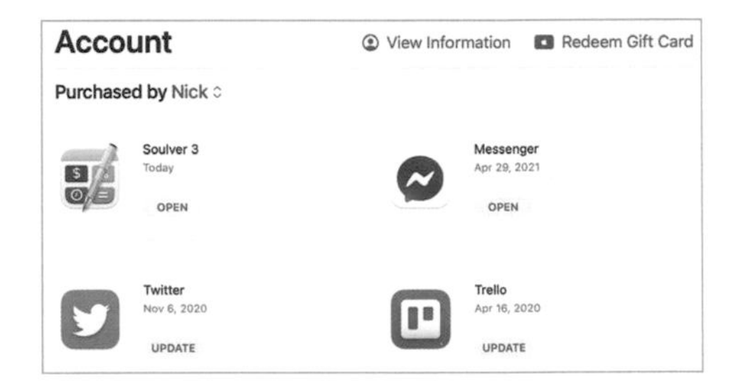

Updating apps

Improvements and fixes are being developed constantly, and these can be downloaded to ensure that all of your apps are up-to-date.

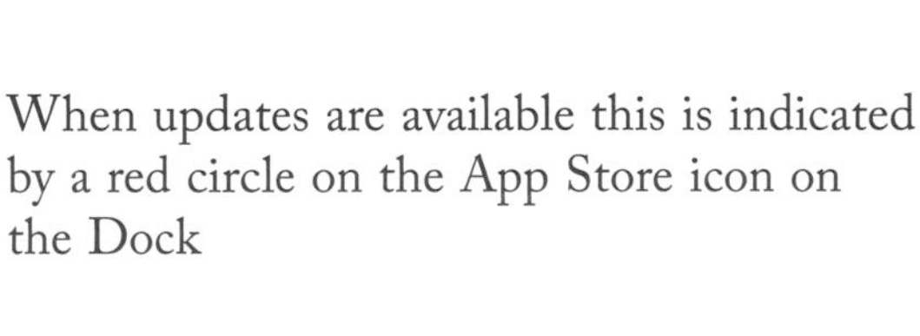

1. When updates are available this is indicated by a red circle on the App Store icon on the Dock

Apps can also be updated automatically. This can be specified in **System Preferences** > **Software Update** > **Advanced**. Check **On** the **Download new updates when available:** option. Underneath this there are options for downloading and installing updates.

2. Click on the **Updates** button

Updates

3. Information about the update is displayed next to the app that is due to be updated

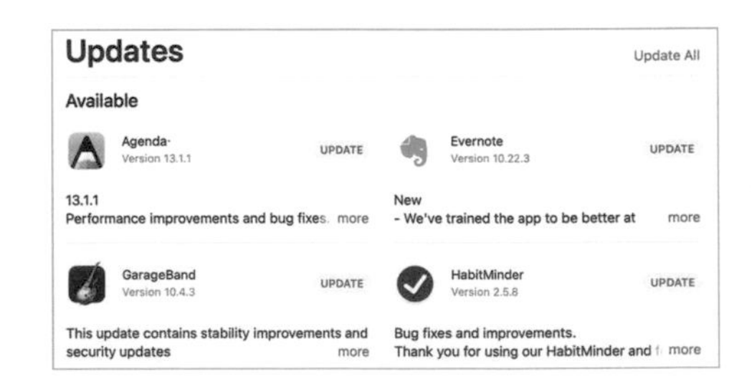

4. Click on the **Update** button to update an individual app

5. Click on the **Update All** button to update all of the apps that are due to be updated

6 Getting Productive

There are several built-in apps within macOS that can be used to create, store and display information for use every day. This chapter shows how to access and use these apps, so that you can make the most of macOS as an effective and efficient productivity tool for a variety of daily tasks.

Contacts (Address Book)

The Contacts app can be used to store contact information, which can then be used in different apps and shared via iCloud. To view contacts:

Contacts can be shared by clicking on the **Share** button and selecting one of the available options.

Email Card
Message Card
AirDrop Card
Notes
More...

Open **Contacts** from this address book icon, and click on one of the contacts in the left-hand panel to view their details

Adding contact information

The main function of the Contacts app is to store details of personal and business contacts. Contacts must be added manually for each entry, but it can prove to be a valuable resource once this has been completed. To add contact information:

Click on the **Edit** button to edit a contact's details.

Edit

Click on this button to add a new contact

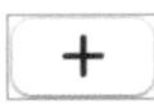

2 Click on a category, and enter contact information. Press the **Tab** key to move to the next field

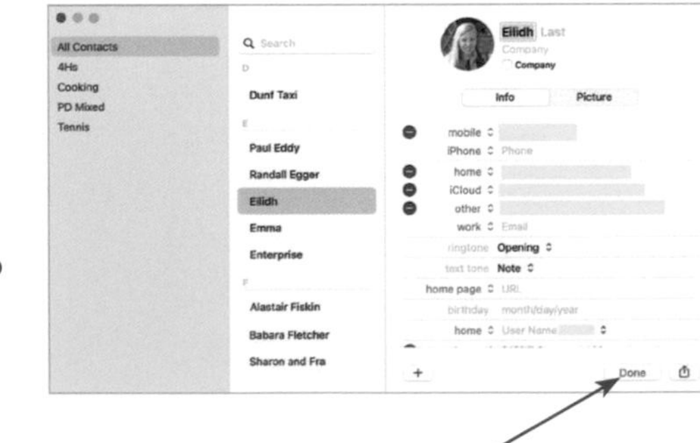

Click on the **Done** button once you have completed the entry

Creating groups

In addition to creating individual entries in the Contacts app, group contacts can also be created. This is a way of grouping contacts with similar interests or connections. Once a group has been created, all of the entries within it can be accessed and contacted by selecting the relevant entry in the left-hand panel. To create a group:

Individuals can be included in several groups. If you change their details in one group, these changes will take effect across all of the groups in which the entry occurs.

1. Select **File** > **New Group** from the Menu bar to create a new group entry

2. Give the new group a name

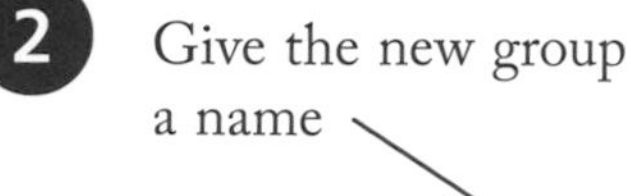

3. Drag individual entries into the group (the individual entries are retained, too)

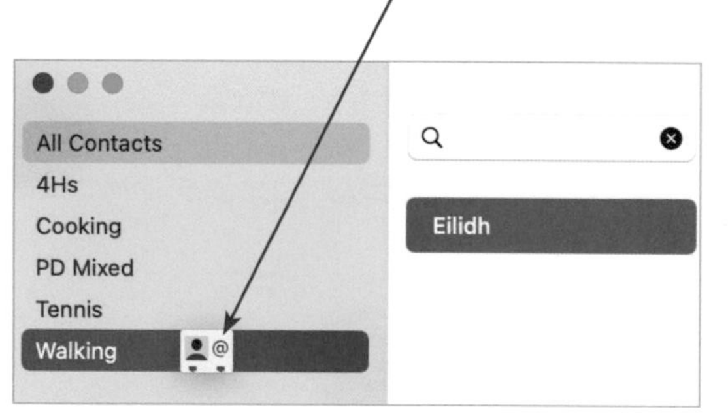

4. Click on a group name to view the members of the group

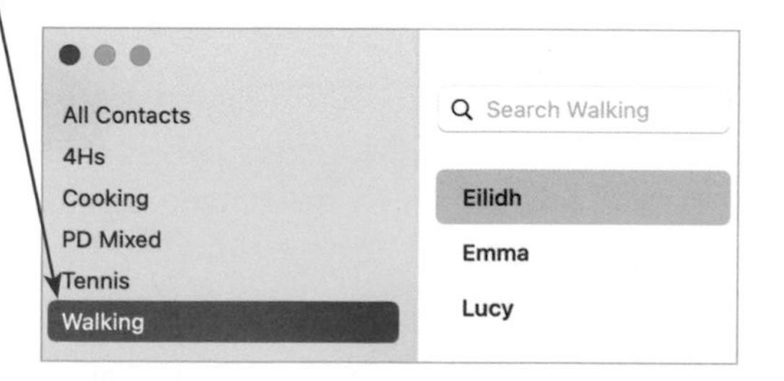

Groups in the Contacts app can be used to send group emails; i.e. you can type the name of a group into the **Mail To:** box to generate names in the group, and then send an email to all of these recipients.

Calendar

Electronic calendars are now a standard part of modern life, and with macOS this function is performed by the Calendar app. Not only can this be used on your Mac; it can also be synchronized with other Apple devices such as an iPad or an iPhone, using iCloud. To create a calendar:

Hot tip

The Calendar app displays the current date in the icon on the Dock.

1. Click on this icon on the Dock, or in the Launchpad

2. Select whether to view the calendar by Day, Week, Month or Year

Don't forget

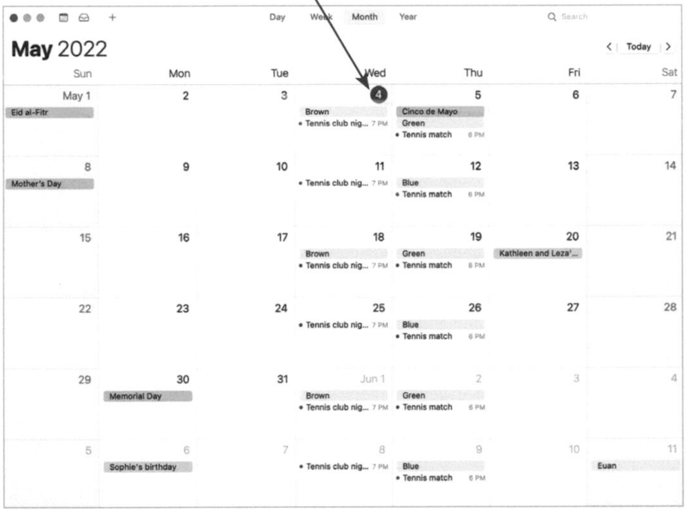

Click on the **Today** button to view the current day. Click on the Forward or Back arrows to move to the next day, week, month or year, depending on what is selected in Step 2.

3. In Month view, the current day is denoted by a red circle on the date

Hot tip

If Family Sharing has been activated, the Family calendar will automatically be added for all members of the Family Sharing group.

4. Scroll up and down to move through the weeks and months. In macOS Monterey this is done with continuous scrolling, which means you can view weeks across different months; i.e. you can view the second half of one month and the first half of the next one in the same window

Adding events

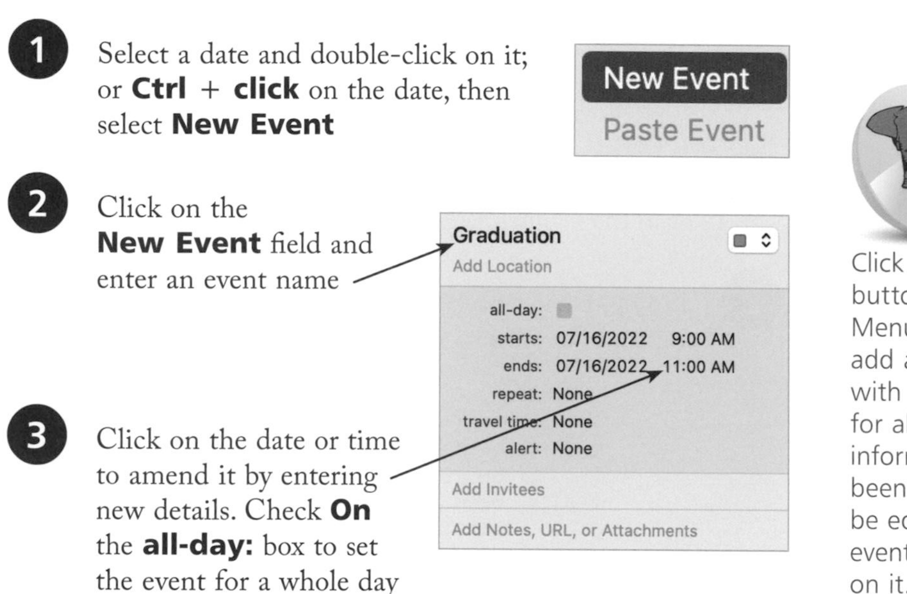

1. Select a date and double-click on it; or **Ctrl** + **click** on the date, then select **New Event**
2. Click on the **New Event** field and enter an event name
3. Click on the date or time to amend it by entering new details. Check **On** the **all-day:** box to set the event for a whole day

Don't forget

Click on this button on the Menu bar to add a **Quick Event**, with just one text box for all of the relevant information. Once it has been added it can then be edited like a regular event by double-clicking on it.

Finding locations

When adding events you can also find details about locations.

1. Click on the **Add Location** field, and start typing a destination name or zip (postal) code. Suggestions will appear underneath, including matching items from your contacts list. Click on a location to select it

Graduation
Add Location
Jul 16, 2022 9 AM to 11 AM
Add Invitees
Add Notes, URL, or Attachments

2. A map of the location is displayed, including a real-time weather summary for the location. Click on the map to view it in greater detail in the **Maps** app

Graduation
McEwan Hall
George Square area, Bristo Square, Edinburgh, Scotland
Jul 16, 2022 9 AM to 11 AM
Alert when I need to leave
Add Invitees
Add Notes, URL, or Attachments
The Three Sisters
DUMBIEDYKES
TOLLCROSS
Edinburgh, Scotland — H:64° L:50°

Don't forget

Other items that can be included in an event are: inviting people to it; calculating the travel time from your current location; creating a repeat event; and setting an alert for it.

Taking Notes

It is always useful to have a quick way of making notes of everyday things such as shopping lists, recipes or packing lists for traveling. With macOS Monterey, the Notes app is perfect for this task. To use it:

Enable iCloud for Notes so that all of your notes will be backed up, and also available on all of your iCloud-enabled devices. You will also be able to access them by signing in to your account with your Apple ID at www.icloud.com

1. Click on this icon on the Dock, or in the Launchpad

2. The right-hand panel is where notes are created. The middle panel displays a list of all notes

Don't forget

The first line of a note becomes its heading in the notes panel.

3. Click on this button on the top toolbar to view the Notes Homepage as a gallery of thumbnails, rather than the default list option. This makes it easier to identify specific lists and their content

4. Click on this button to add a new note

Hot tip

Ctrl + **click** on a note in the middle panel and click on the **Pin Note** option, to pin the note at the top of the panel.

5. As more notes are added, the most recent appears at the top of the list in the middle panel (below any notes that have been pinned). Double-click the new note to add text and edit formatting options (see the next page)

Formatting notes

In macOS Monterey there are a number of formatting options for the Notes app.

1. Enter a line of text, and click on this button to add a check button

2. Click on the check button to add a check mark, to indicate that an item has been completed

3. Highlight a piece of text, and click on this button to access formatting options for it

4. Click on this button on the Notes toolbar to add photos and videos, sketches, scans, maps, websites, sound clips and documents to a note

Click on this button to add a table to a note:

Click on the **Share** button to share a note with other people, via a range of other apps.

Click on this button to invite other people to have access to the note so they can view it and edit it:

Content can be added to notes from a range of other apps, such as Safari or the Photos app, by clicking on the **Share** button in the relevant app and selecting **Notes** as the option.

Quick Notes

The functionality of the Notes app has been expanded with macOS Monterey, so that notes can be created from within any other app without having to open the Notes app separately. This is known as Quick Notes. Once these have been created, they are all stored in the Notes app. To create and manage Quick Notes:

Quick Notes is a new feature in macOS Monterey.

From within any app, move the cursor over the bottom right-hand corner of the screen and click on the white square

2 The Quick Notes panel appears

The title of a Quick Note is taken from the first line of text. If no text is entered, the title will be **New Note**.

3 Enter text for the Quick Note, as required

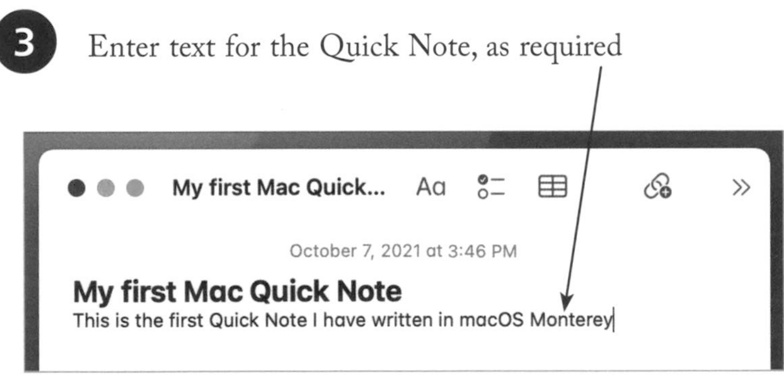

When creating a Quick Note, use these buttons at the top of the window to, from left to right: format selected text in the Quick Note; add checklists to lines of text; and add a table to the Quick Note

All of the notes in the Notes app can be viewed in Activity view, where you can see content that has been added by other people to any notes that you have shared, using the Share button as shown on page 107. To access Activity view, select **View** > **Show Note Activity** from the Menu bar. This is a new feature in macOS Monterey.

5. Click on the **Notes** app to open it

6. The Quick Note is available in the left-hand sidebar and can be opened in the same way as a regular note; i.e. by clicking on it

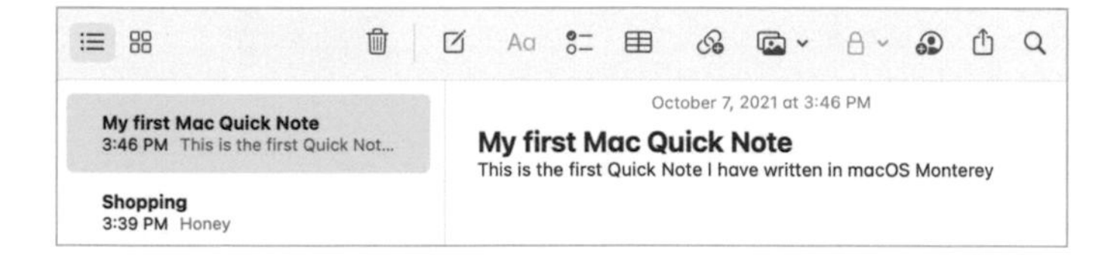

Adding links

Links can be added to Quick Notes, from the currently-active app, for some apps. One options is to add a link to a web page.

1. Open a web page in the Safari app and create a new Quick Note, as shown on the previous page. Click on this button to add a link to the web page currently being viewed

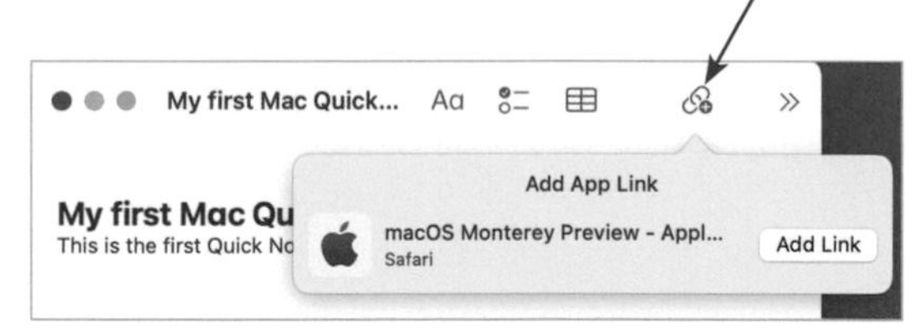

2. The web page link is added to the Quick Note and this can be used to access the web page; see the Hot tip

Once content has been added to a Quick Note, it can then be accessed from the Quick Note in the Notes app; i.e. if a web link has been added, click on the link in the Quick Note to go to the linked web page.

Setting Reminders

Another useful app for keeping organized is Reminders. This enables you to create lists for different topics and also set reminders for specific items. A date and time can be set for each reminder, and when this is reached, the reminder appears on your Mac screen (and in the Notification Center). To use Reminders:

Click on this icon on the Dock, or in the Launchpad

The Reminders window contains a **My Lists** section where lists and reminders are created and a Smart Lists section, at the top of the left-hand sidebar, where matching items are collated

Click on the one of the **My Lists** options; e.g. **Reminders**

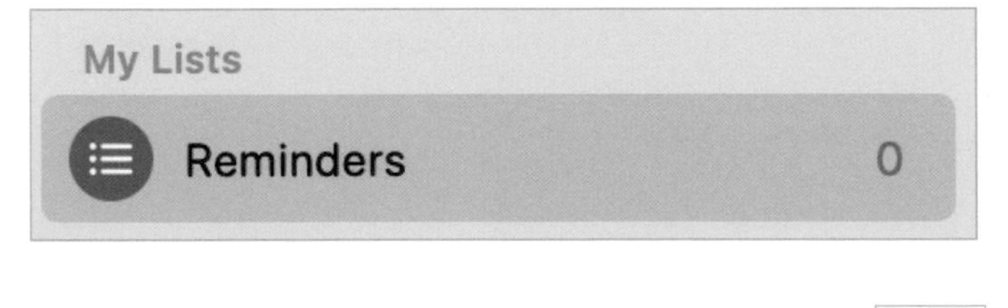

Click on this button to add a new reminder

Don't forget

As with Notes, iCloud makes your reminders available on all of your Apple devices; i.e. your Mac, iPad and iPhone.

5. Enter the title of the reminder and click on the **i** button to access the **Details** window

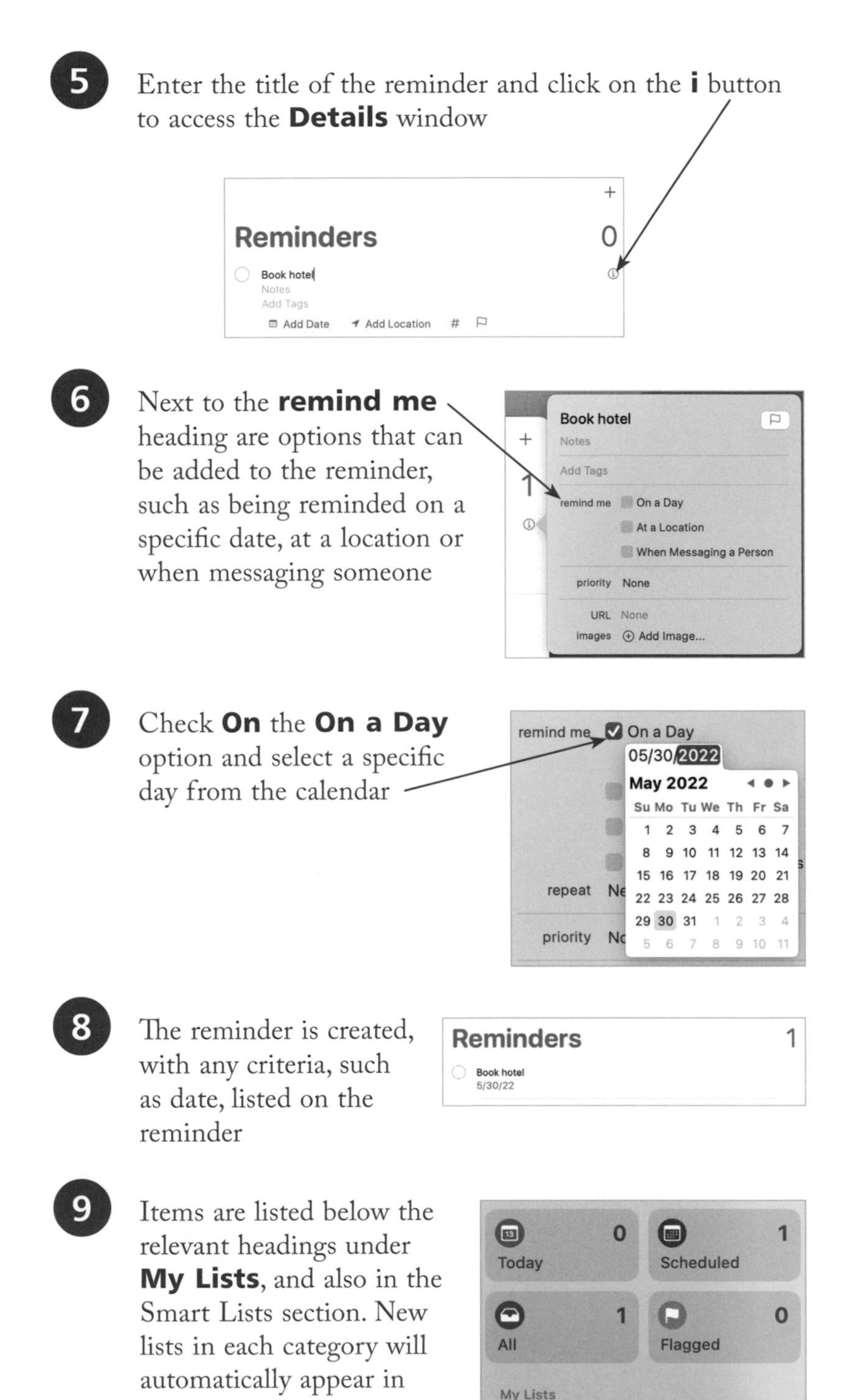

6. Next to the **remind me** heading are options that can be added to the reminder, such as being reminded on a specific date, at a location or when messaging someone

7. Check **On** the **On a Day** option and select a specific day from the calendar

8. The reminder is created, with any criteria, such as date, listed on the reminder

9. Items are listed below the relevant headings under **My Lists**, and also in the Smart Lists section. New lists in each category will automatically appear in each section

Hot tip

If the **On a Day** option is selected in Step 7, a **repeat** option then becomes available. This can be used to create a recurring reminder. The options for this are **None**, **Every Day**, **Every Week**, **Every Month**, **Every Year**, or **Custom**.

News

The News app has been a regular feature on Apple's mobile devices for a number of years. It is also available on Macs using macOS Monterey. To use it:

1. Click on the **News** app

2. Click on the **Today** button in the left-hand panel to view the latest stories in your news feed

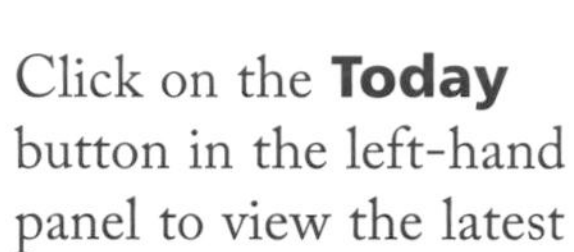

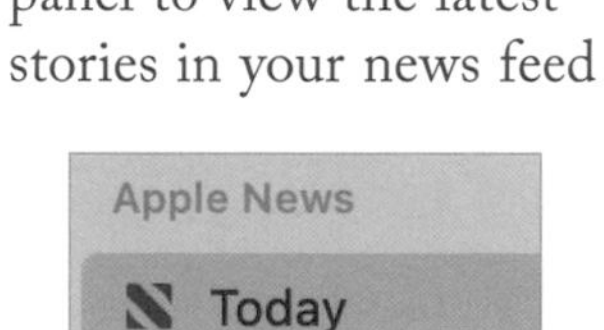

Click on the **News+** button in the left-hand sidebar to access the subscription service for Apple News.

3. On the News app Menu bar, select **File** > **Discover Channels...** to view subjects or publications that you are following; i.e. they are used to populate your news feed

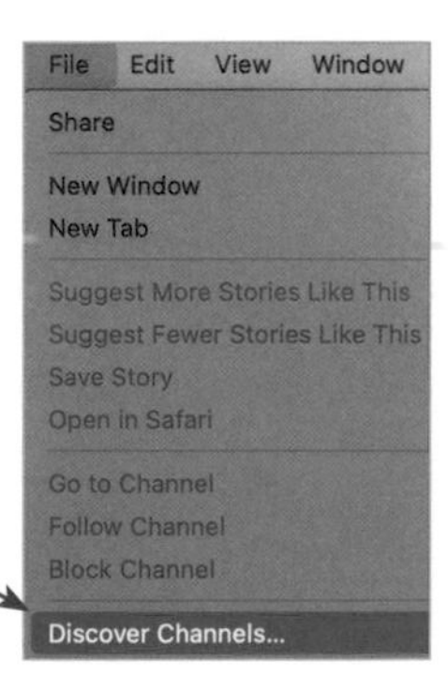

4. In the **Channels** section, click on items to select them and add them as topics or publications that are added as favorites on your Today page

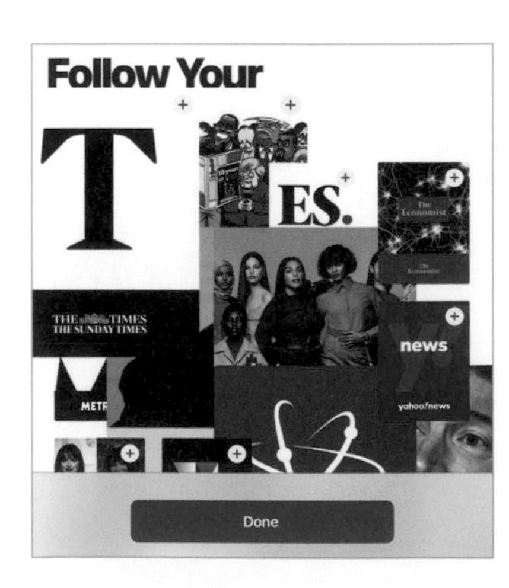

Voice Memos

Voice memos are an increasingly popular way to send short voice messages to family and friends. They can also be used as a method of verbal note-taking, when you want to remember something. To use voice memos:

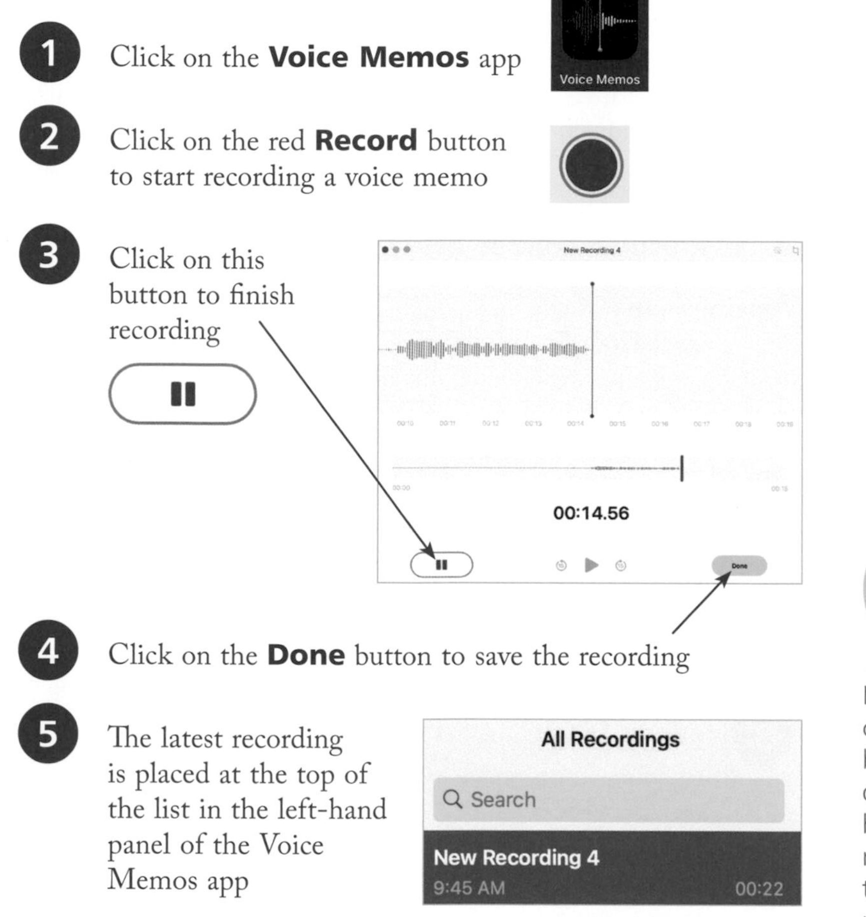

1. Click on the **Voice Memos** app
2. Click on the red **Record** button to start recording a voice memo
3. Click on this button to finish recording
4. Click on the **Done** button to save the recording
5. The latest recording is placed at the top of the list in the left-hand panel of the Voice Memos app
6. Click on a memo to display it in the main window. Click on the **Play** button to hear the recording
7. Click on the **Edit** button in the top right-hand corner of the main window to amend a voice memo

In the **Edit** window, click on the **Crop** button and drag the yellow handles left or right to trim the duration of a voice memo. Tap on the **Trim** button below the yellow handles to complete the operation. This is a good option if there are pauses at the beginning or end of the recording.

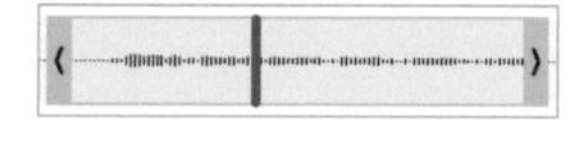

Getting Around with Maps

With the Maps app you need never again wonder about where a location is, or worry about getting directions to somewhere. Using Maps with macOS Monterey, you will be able to do the following:

- Search maps from around the world
- Find addresses
- Find famous buildings or landmarks
- Get directions between different locations
- View transit information for a route
- View traffic conditions

The Maps app has been updated in macOS Monterey, with graphics-rich examples of cities in Explore and Satellite views.

Viewing maps

Enable **Location Services** and then you can start looking around maps, from the viewpoint of your current location.

Location Services can be enabled in **System Preferences** > **Security & Privacy**. Click on the **Privacy** tab, click on **Location Services** and check **On** the **Enable Location Services** checkbox. (You may need to click the lock in the bottom-left corner and enter your sign-in password before you can make any changes.) Maps can be used without Location Services but this would mean that Maps cannot use your current location or determine information in relation to this.

Click on this button on the Dock or in the Launchpad

Click on this button to view your current location

3 Double-click to zoom in on the map. Press **Option + double-click** to zoom out. Or, swipe outward with thumb and forefinger to zoom in, and pinch inward to zoom out

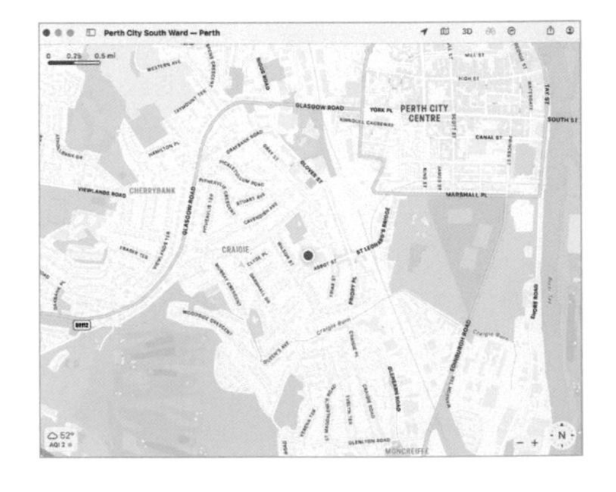

Click on this button to view, or hide, the left-hand sidebar

Finding locations

Locations in Maps can be found for addresses, cities or landmarks. To find items in Maps:

1. Enter a term into the Search box and click on one of the results

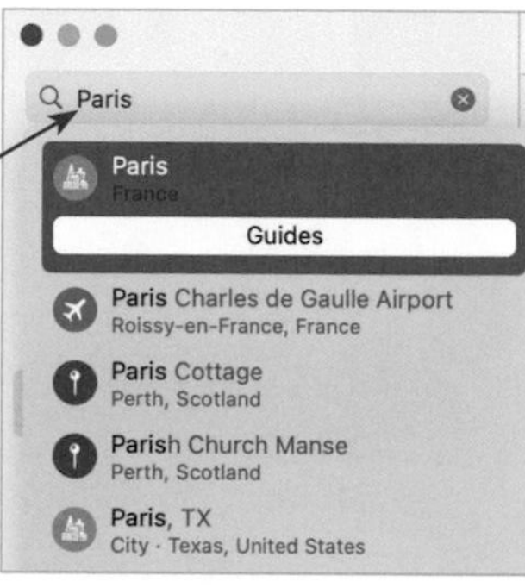

Flyover Tour (if available for a location) provides an animated 3D experience that travels through the most notable sights for that location.

2. The selected item is displayed and shown on a map with a panel displaying information about it. You can also access a **Flyover Tour** of the location (if available)

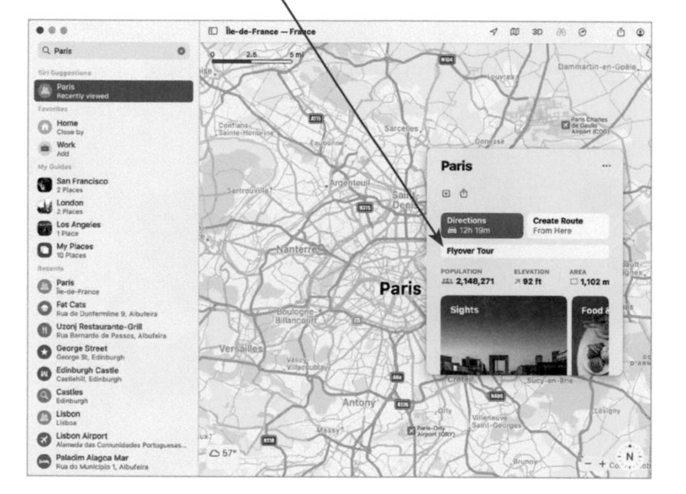

3. Click on this button on the top toolbar to select **Explore**, **Driving**, **Transit** or **Satellite** views of the location

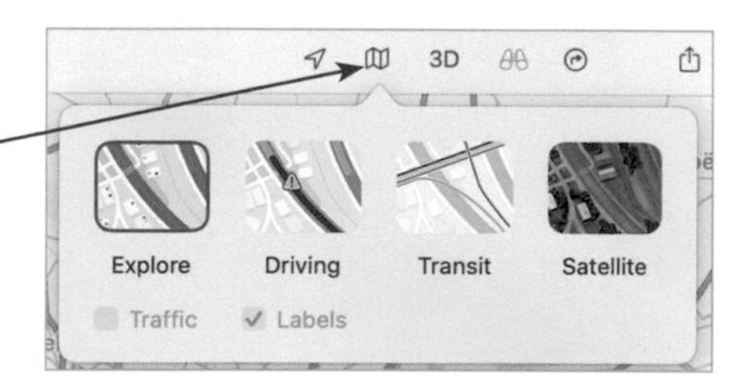

Click on the 3D button on the top toolbar to change the perspective of the map being viewed. Rotate your fingers on the trackpad to change the orientation of the map.

Getting directions

Within Maps you can also get directions to most locations.

Click on this button in Step 2 to swap the locations for which you want directions:

Search for a location and click on the **Directions** button

By default, your current location is used for the **Start** field. If you want to change this, click once and enter a new location

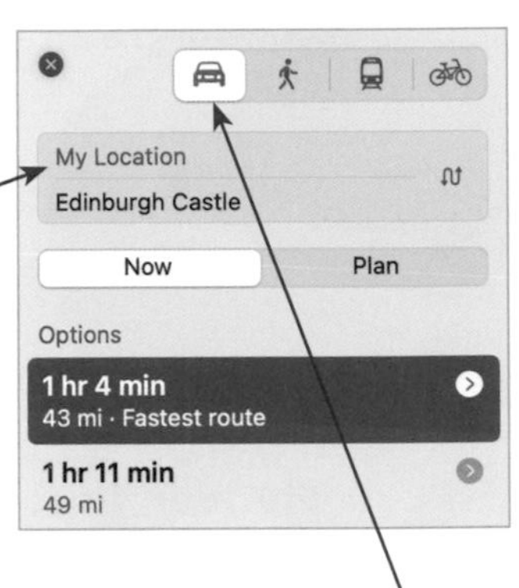

Click on one of the options for reaching your destination (**Drive**, **Walk**, **Transit** or **Cycle**). The route is shown on the map with the directions. The default mode of transport is for driving

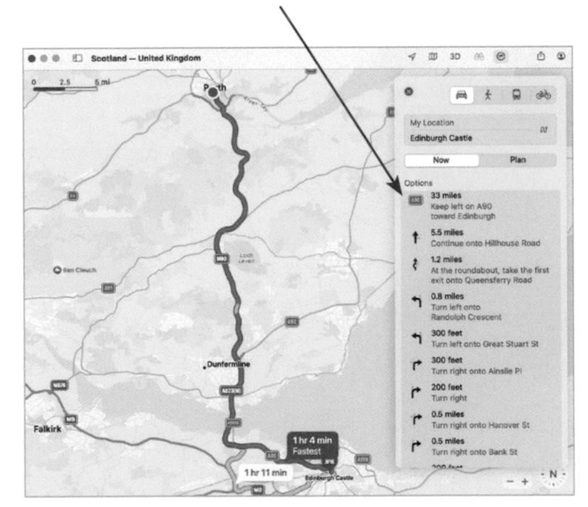

Click on the **Share** button on the top toolbar to send the directions to a mobile device, such as an iPhone or an iPad, so that you can follow the directions on the go.

Select the options in Step 3 on page 115 and click on the **Driving** button to view traffic conditions for the selected route

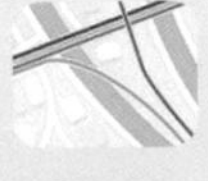

Finding transit directions

In macOS Monterey there is an option for accessing transit options between two locations. To do this:

1. Enter the two locations, as shown on the previous page. Click on the **Transit** button

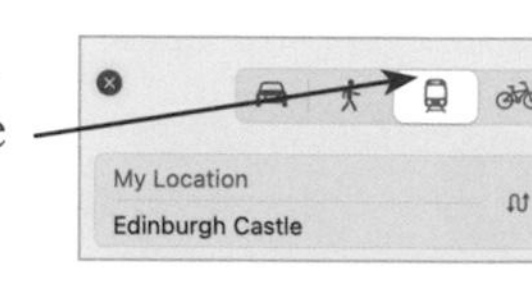

2. The transit options are shown in a panel on the map. Click here next to an option to view its full details. This will include any parts of the journey where you will have to walk between the transit services required for your journey

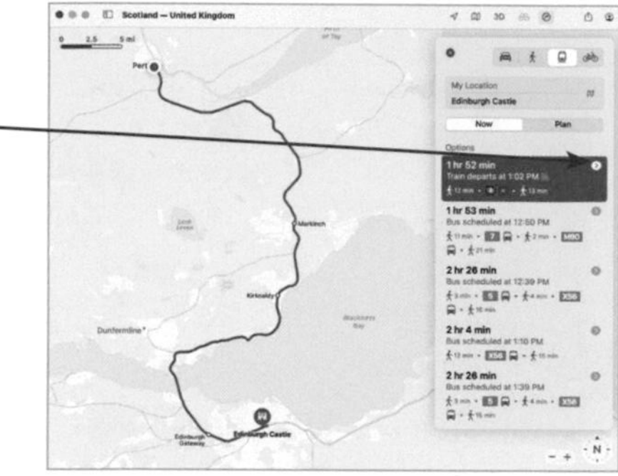

Click on the **Plan** button in Step 2 to view the transit details for a specific date and time.

Now Plan
Leave Arrive
6/14/2022 2:58 PM

Look Around

If a specific street is being viewed, there is an option to view it with a photographic interface, from where you can explore the location and move around the area. To do this:

1. Access a street location and click on this icon on the top toolbar

2. The **Look Around** window is activated over the map. Click within the Look Around window to move through the location, or click on this icon to move its location on the map

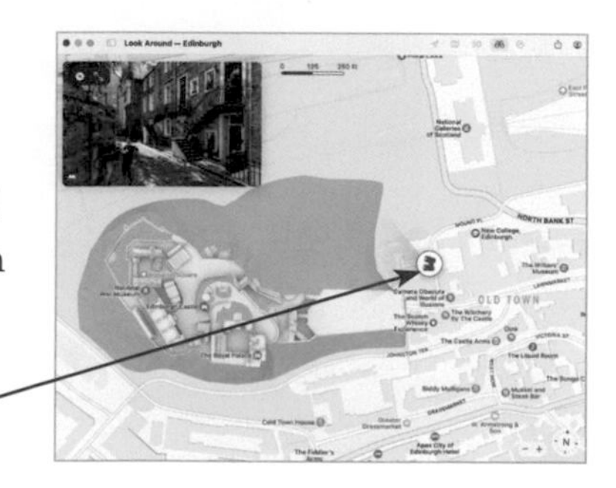

The Transit option is only available for a limited number of locations around the world – the majority being in the US.

Preview

Preview is a macOS app that can be used to view multiple file types, particularly image file formats. This can be useful if you just want to view documents without editing them in a dedicated app, such as an image-editing app. Preview in macOS Monterey can also be used to store and view documents in iCloud.

To open Preview, click on this icon in the Applications folder in the Finder, on the Dock or in the Launchpad:

Open **Preview** and click on one of the options in the Finder sidebar

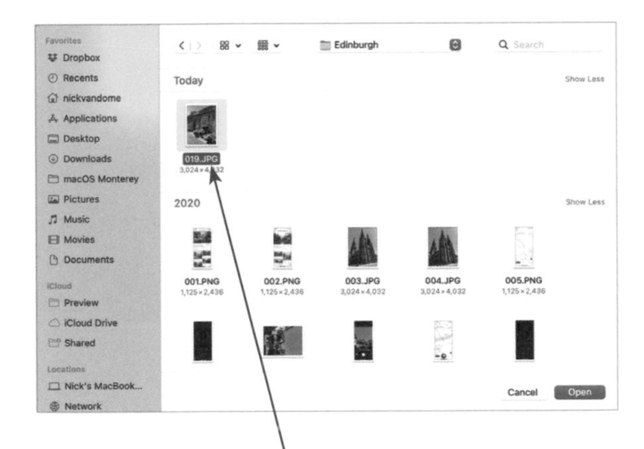

Double-click on an item to view it at full size. This can be from any of the folders within the Finder. The item is opened in the Preview window. Use the Preview toolbar to manage and edit open items

Preview is also a good option for viewing PDF (Portable Document Format) documents.

Printing

macOS Monterey makes the printing process as simple as possible, partly by its ability to automatically install new printers as soon as they are connected to your Mac. However, it is also possible to install printers manually. To do this:

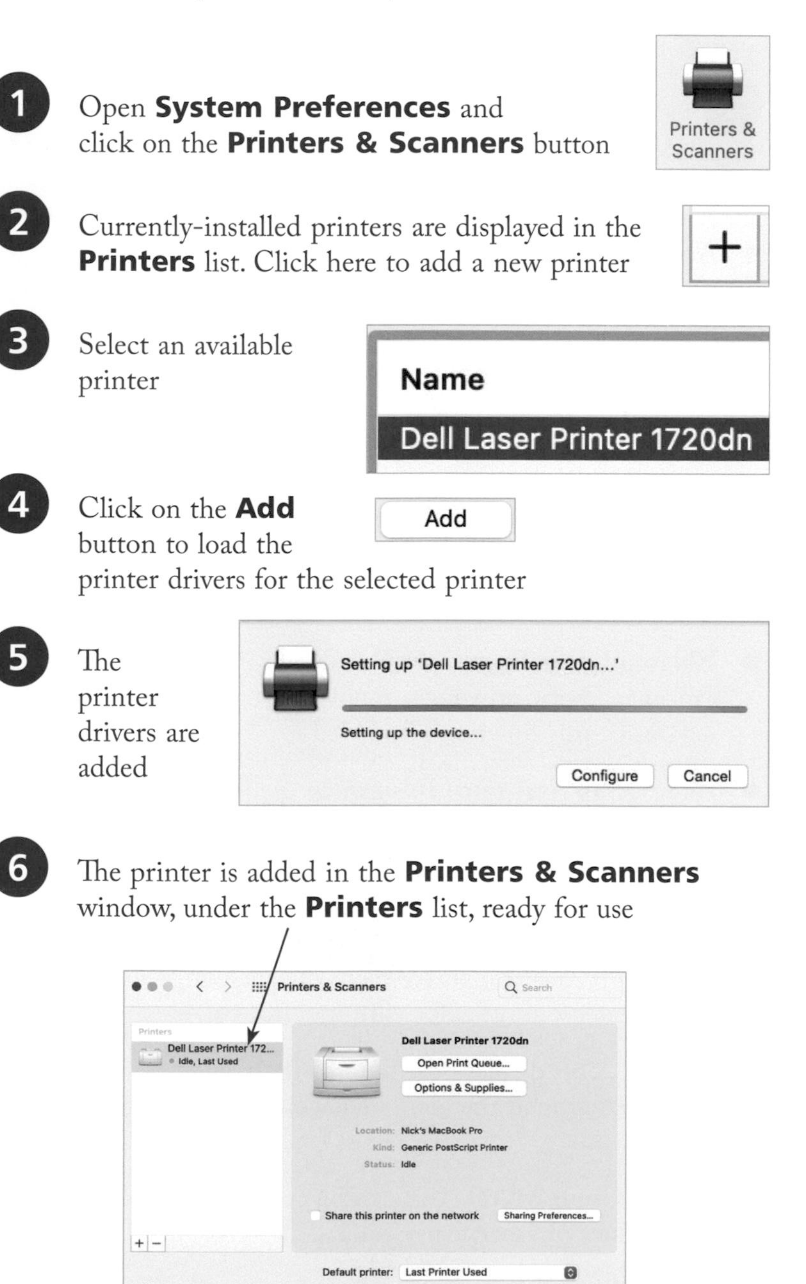

1. Open **System Preferences** and click on the **Printers & Scanners** button
2. Currently-installed printers are displayed in the **Printers** list. Click here to add a new printer
3. Select an available printer
4. Click on the **Add** button to load the printer drivers for the selected printer
5. The printer drivers are added
6. The printer is added in the **Printers & Scanners** window, under the **Printers** list, ready for use

For most printers, macOS will detect them when they are first connected, and they should be ready to use without the need to install any software or apply new settings.

Once a printer has been installed, documents can be printed by selecting **File** > **Print** from the Menu bar. Print settings can be set at this point, and they can also be set by selecting **File** > **Page/Print Setup** from the Menu bar in most apps.

macOS Utilities

In addition to the apps in the Applications folder, there are also a number of utility apps that perform a variety of tasks within macOS. (Some of the utilities vary depending on the hardware setup of your Mac.) The utility apps are in the **Other** folder.

The Utilities folder is also available from within the Applications folder. On some systems it may appear as "Other" in the Launchpad.

1. Access the Launchpad to access the **Other** folder. The utilities are displayed within the Other folder

- **Activity Monitor**. This contains information about the system memory being used and disk activity (see page 180 for more details).
- **AirPort Utility**. This sets up the AirPort wireless networking facility that can be used to connect to the internet with a Wi-Fi connection (pre-2018).
- **Audio MIDI Setup**. This can be used for adding audio devices and setting their properties.
- **Automator**. This can be used to create automated tasks.
- **Bluetooth File Exchange**. This determines how files are exchanged between your computer and other Bluetooth devices (if this function is enabled).
- **Boot Camp Assistant**. This can be used to run Windows operating systems on your Mac.
- **Chess**. A game that can be played against your Mac.
- **ColorSync Utility**. This can be used to view and create color profiles on your computer. These can then be used by apps to try to match output color with monitor color.
- **Console**. This displays the behind-the-scenes messages that are passed around the computer while its usual tasks are being performed.
- **Digital Color Meter**. This can be used to measure the exact color values of a particular color.
- **Disk Utility**. This can be used to view details about attached disks and repair errors (see page 178 for more details).

The utilities are the workhorses of macOS. They do not have the glamour of apps such as the Music app, but they perform vital gathering of information and general maintenance tasks.

You may never need to use a utility like the Console, but it is worth having a look at it just to see the inner workings of a computer.

- **Feedback Assistant**. This can be used to receive and send feedback updates about Apple and macOS.
- **Font Book**. This displays the available system fonts.
- **Grapher**. This is a utility for creating simple or more complex scientific graphs.
- **Image Capture**. This can be used to capture images, or transfer and scan images.
- **Keychain Access**. This can be used to manage passwords for websites and also create new secure passwords as required.
- **Migration Assistant**. This helps in the transfer of files between two Mac computers. This can be used if you buy a new Mac and you need to transfer files from another one.
- **QuickTime Player**. This can be used to play video files.
- **Screenshot**. This is a utility that can be used to capture screenshots, which are images of the screen at a given point in time. You can grab different portions of the screen, including timed options.
- **Script Editor**. This can be used to create your own scripts with Apple's dedicated scripting app, AppleScript.
- **Shortcuts**. This can be used to create automated actions.
- **Stickies**. This can be used to attach short notes to the screen.
- **System Information**. This contains details of the hardware devices and software applications that are installed on your computer (see page 179 for more details).
- **Terminal**. Within the Terminal you can view the workings of UNIX and also start to write your own apps, if you have some UNIX programming knowledge.
- **TextEdit**. This is a basic word processing app.
- **Time Machine**. This is the macOS backup facility.
- **VoiceOver Utility**. This has various options for the digital voice that can be used to read out what is on the screen, and it is particularly useful for users who are visually impaired.

The Screenshot utility is a feature that is useful if you are producing manuals or books, and need to display examples of a screen or app.

Creating PDF Documents

PDF (Portable Document Format) is a file format that preserves the formatting of an original document, and it can also be viewed on a variety of computer platforms including Mac, Windows and UNIX. macOS has a built-in PDF function that can produce PDF files from many apps. To use this:

PDF files can be viewed with the Preview app.

1. Open a file in an app and select **File** > **Export as PDF...**, or use the **File** > **Save As** option and change the format of the document to PDF, where offered

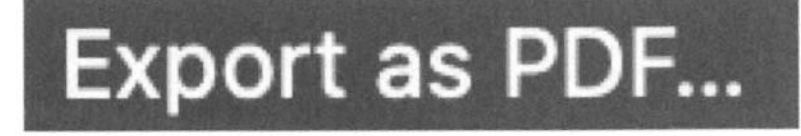

2. Browse to a destination for the file, and click **Save**

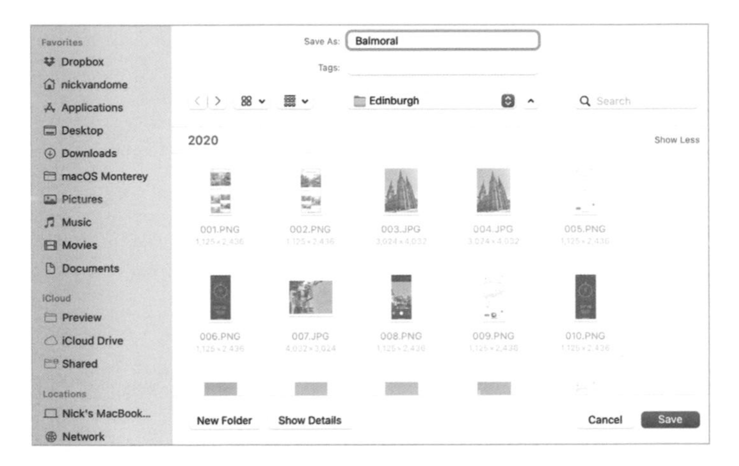

PDF is an excellent option if you are creating documents such as instruction booklets, magazines or manuals.

3. Locate the selected destination to view and open the newly-created PDF file. You can then access options for viewing, sharing or marking up the document

7 Internet and Communication

This chapter shows how to get the most out of the internet and keeping in touch. It covers how to use the macOS web browser, Safari, and its email app, Mail. It also covers Messages for text messaging, and video calling with the updated FaceTime app.

About Safari

Safari is a web browser that is designed specifically to be used with macOS. It is similar in most respects to other browsers, but it usually functions more quickly with macOS.

Before you connect to the internet you must have an Internet Service Provider (ISP) who will provide you with the relevant method of connection; e.g. cable, wireless or dial-up (if being used). They will also provide you with any passwords (for connecting to a Wi-Fi router) and login details. Click on the **Network** option in System Preferences to view the available networks.

Safari overview

1. Click here on the Dock to launch Safari

2. All of the controls are at the top of the browser

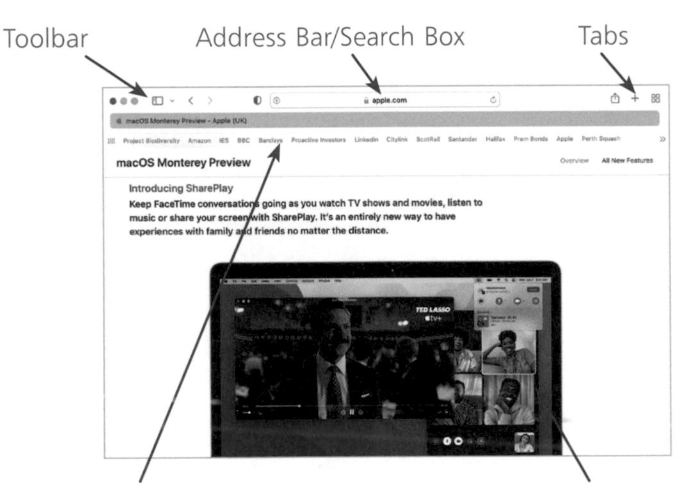

If the Favorites Bar is not visible, select **View** from the Menu bar and check on the **Show Favorites Bar** option. From this menu you can also select or deselect items such as showing the sidebar, the Status Bar and the Reading List. You can also display the sidebar by clicking on this button:

Address Bar/Search Box

In Safari, the Address Bar and the Search Box are combined, and can be used for searching for an item or entering a web address to go to that page.

1. Click in the Address Bar/Search Box

2. Type a website address or a keyword. As you type, suggestions appear below the Address Bar/Search Box. Click on the **Top Hit** option to go to that website

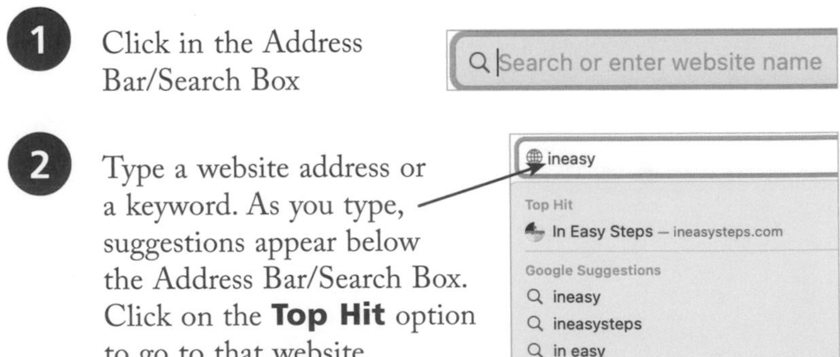

Start Page

When you first open Safari or open a new tab, the Start page is displayed. This can contain a selection of items that can be set within the Safari preferences. The options include favorites, frequently-visited pages and items selected by Siri, but the items displayed can be modified (see page 126). To use the Start page:

1 The Start page is displayed when Safari is opened or when a new tab is opened (unless specified differently; see the second Hot tip on page 127)

Swipe up the Start page to view the available options

Safari is a full-screen app and can be expanded by clicking the green button in the top left-hand corner. For more information on full-screen apps, see pages 92-93.

Move the cursor over the Favorites section on the Start page and click on the **Show All** button to view more favorite pages. Move the cursor over the Favorites section again and click on the **Show Less** button to reduce the number of visible items.

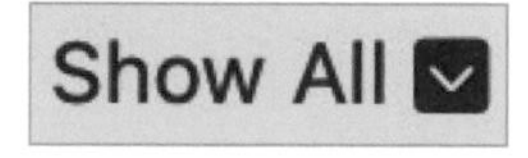

...cont'd

Ctrl + **click** anywhere on the Start page to access the menu for displaying items. Check items **On** or **Off** as required

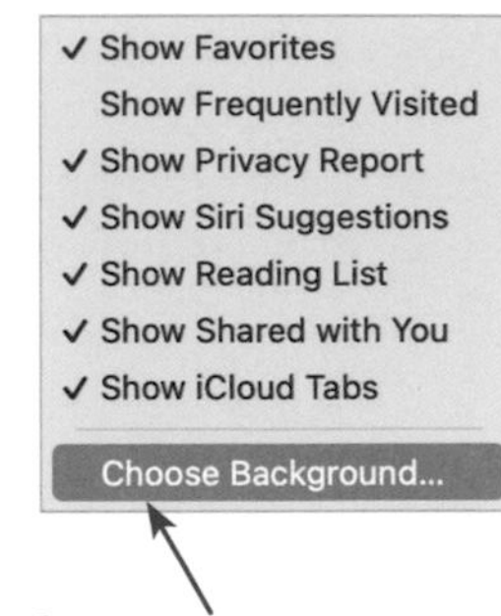

The **Show iCloud Tabs** option for the Start page displays open tabs in Safari from other Apple devices, such as an iPhone or an iPad, provided that they are using the same Apple ID.

Click on the **Choose Background...** option to select a background image for the Start page

5. Navigate to an image within the Finder, select it and click on the **Choose** button to add it as a background

The background image is added to the Start page, behind the existing elements on the page

7. One of the options for displaying on the Start page is a **Privacy Report**, which can be used to show certain security issues with pages in Safari

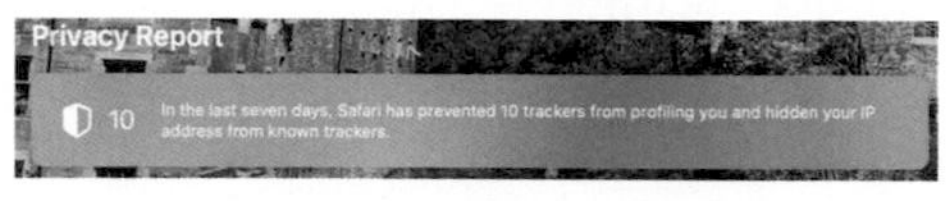

8. Click on the Privacy Report on the Start page to view full details for a 30-day period of using Safari, for elements that track your actions across websites

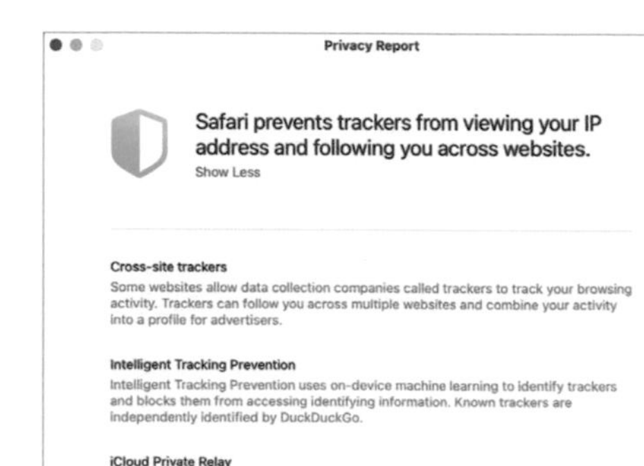

Items that appear in new windows or tabs can be customized with the Safari preferences. To do this:

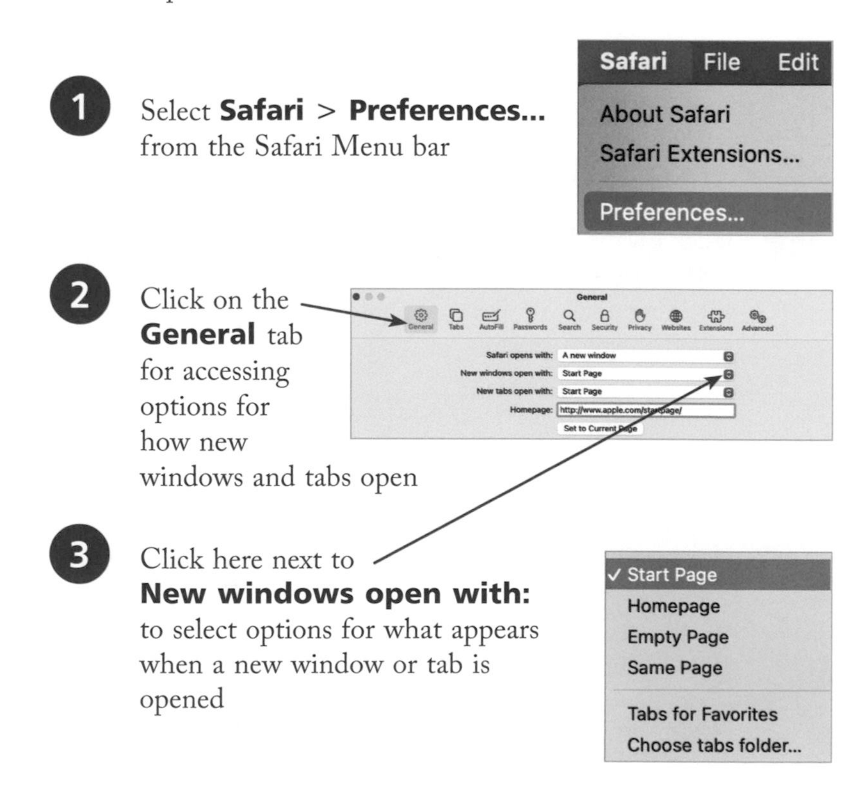

1. Select **Safari** > **Preferences...** from the Safari Menu bar

2. Click on the **General** tab for accessing options for how new windows and tabs open

3. Click here next to **New windows open with:** to select options for what appears when a new window or tab is opened

Privacy Reports can be accessed for individual pages by clicking on this button on the top toolbar:

A specific page (Homepage) can also be set as the option when a new window or a new tab is opened. The Homepage can be set in the **General** Safari preferences from the **Homepage** option.

The **General** Safari preferences can also be used to specify what appears when a new tab is opened, from the **New tabs open with:** option. These are the same options as those in Step 3.

Safari Sidebar

A useful feature in Safari is the Safari sidebar. This is a panel in which you can view all of your bookmarks and Reading List items, which can be read when the Mac is offline.

The sidebar been updated in macOS Monterey, to include Shared with You (see page 137).

1. Select **View** > **Show Sidebar** from the Safari Menu bar, or click on this button on the top toolbar

2. Click on the **Bookmarks** button to view all of the items that you have bookmarked (see page 130)

3. Click on the **Reading List** button to view all of the items that you have added to your Reading List so that they can be read later, even if you are offline and not connected to the internet. These can be added from the **Share** button on the top toolbar

Click on this button in the top right-hand corner of the sidebar to access options for managing groups of tabs. The options include creating a **New Empty Tab Group**, or creating a **New Tab Group with [] Tabs**, which creates a new tab group using the existing open tabs. Different tab groups can be used to open related web pages within the same group, to keep related subjects together. This is a new feature in macOS Monterey.

Safari Tabbed Browsing

Tabs are now a familiar feature on web browsers, so you can have multiple sites open within the same browser window.

Select **Safari** > **Preferences** > **Tabs** from the Menu bar to specify settings for the way tabs in Safari operate. Click on the **Separate** button to display the tabs bar in the traditional format. Click on the **Compact** button to display the tabs separately, above the Favorites bar. This also enables the color of the web page being viewed to extend to the top of the window. This option for tabs is a new feature in macOS Monterey.

When more than one tab is open, the tabs appear at the top of the web pages

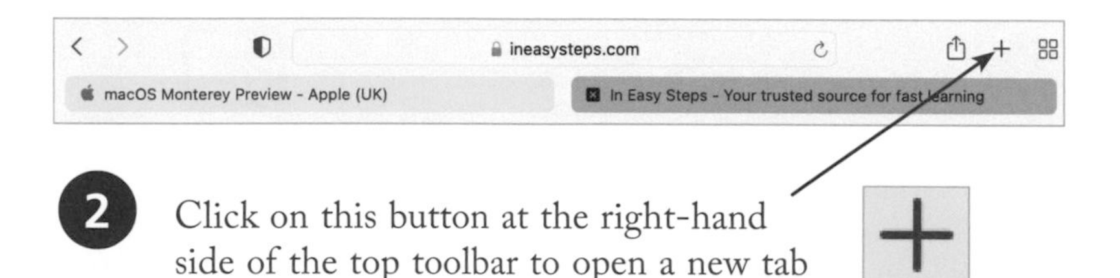

2. Click on this button at the right-hand side of the top toolbar to open a new tab

3. Select one of the options from the Start page, or enter a keyword or phrase, or website address, in the Address Bar/Search Box

4. Click on this button next to the **New Tab** button to minimize all of the current tabs

5. Scroll up and down the window to view all of the open tabs in thumbnail view. Click on one to view it at full size. Click on this button to open a new tab

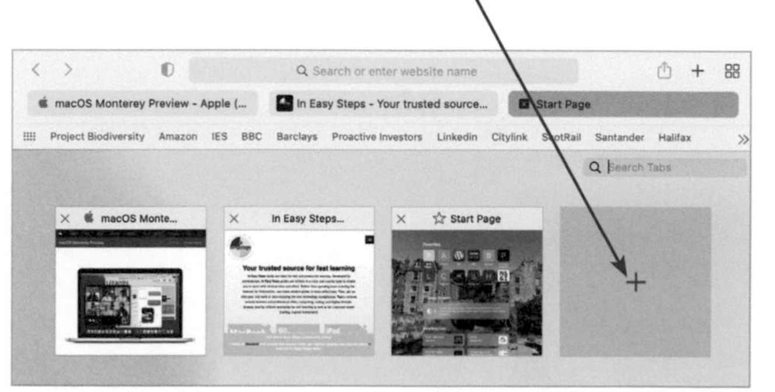

If you have open tabs on other Apple devices (e.g. other Mac computers, iPads or iPhones) these will be listed if you select to open a new tab in Step 5. Click on an item to open it in Safari on your Mac.

Adding Bookmarks

Bookmarks is a feature with which you can create quick links to your favorite web pages or the ones you visit most frequently. Bookmarks can be added to a menu or the Bookmarks panel in Safari, which makes them even quicker to access. Folders can also be created to store less frequently-used bookmarks. To view and create bookmarks:

1. Click on this button to view the sidebar 

2. Access the Bookmarks section as shown on page 128 to view the currently-available bookmarks and any bookmarks folders

3. Click on the **Share** button on the top toolbar, and click on the **Add Bookmark** button to create a bookmark for the page currently being viewed

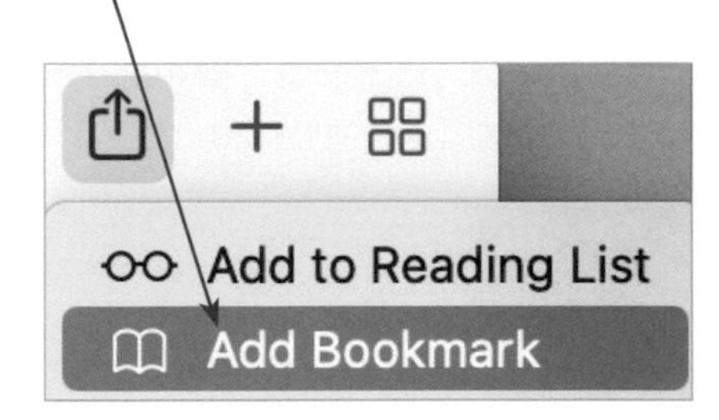

4. Enter a name for the bookmark and select a location for it

5. Click on the **Add** button

Hot tip

Bookmarks can be added to the Favorites Bar, which can be displayed at the top of the Safari window, below the Address Bar. To add bookmarks to the Favorites Bar, select **Favorites** for **Add this page to:** in Step 4. To show (or hide) the Favorites Bar, select **View** > **Show Favorites Bar** (or **Hide Favorites Bar**) from the Safari Menu bar. The Favorites folder is displayed at the top of the Bookmarks panel.

Website Settings

One of the most annoying things about using the web is the adverts that seem to appear by coincidence when you are browsing a website. These are actually based on items that you have previously looked at on the web, and are generated by cookies that have been downloaded from the original site. However, with macOS Monterey a lot of these adverts are blocked automatically, and you can also apply your own settings for specific pages to enable or restrict certain items. To do this:

1. Select **Safari** > **Settings for [website name]** from the Safari Menu bar

2. Make selections as required, such as enabling the Reader function and using content blockers. Click on the **Enable content blockers** option to block unwanted items

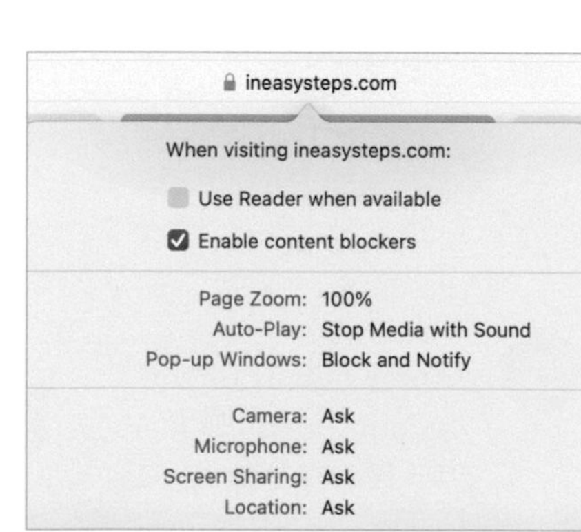

3. Select **Safari** > **Preferences** from the Safari Menu bar, and click on the **Privacy** button on the top toolbar

4. Under **Website tracking:**, check **On** the **Prevent cross-site tracking** checkbox to prevent websites displaying adverts based on what you have viewed on other sites

Don't forget

The Reader function is one where a web page can be read without additional items such as toolbars, buttons and navigation bars. If a page supports the Reader function it will display this button in the Address Bar:

Mail

Email is an essential element for most computer users, and Macs come with their own email app called Mail. This covers all of the email functionality that most people could need.

When first using Mail, you have to set up your email account. This can be done with most email accounts and also a wide range of webmail accounts, including iCloud. To add email accounts:

Don't forget

Mail is a full-screen app. For more information on full-screen apps, see pages 92-93.

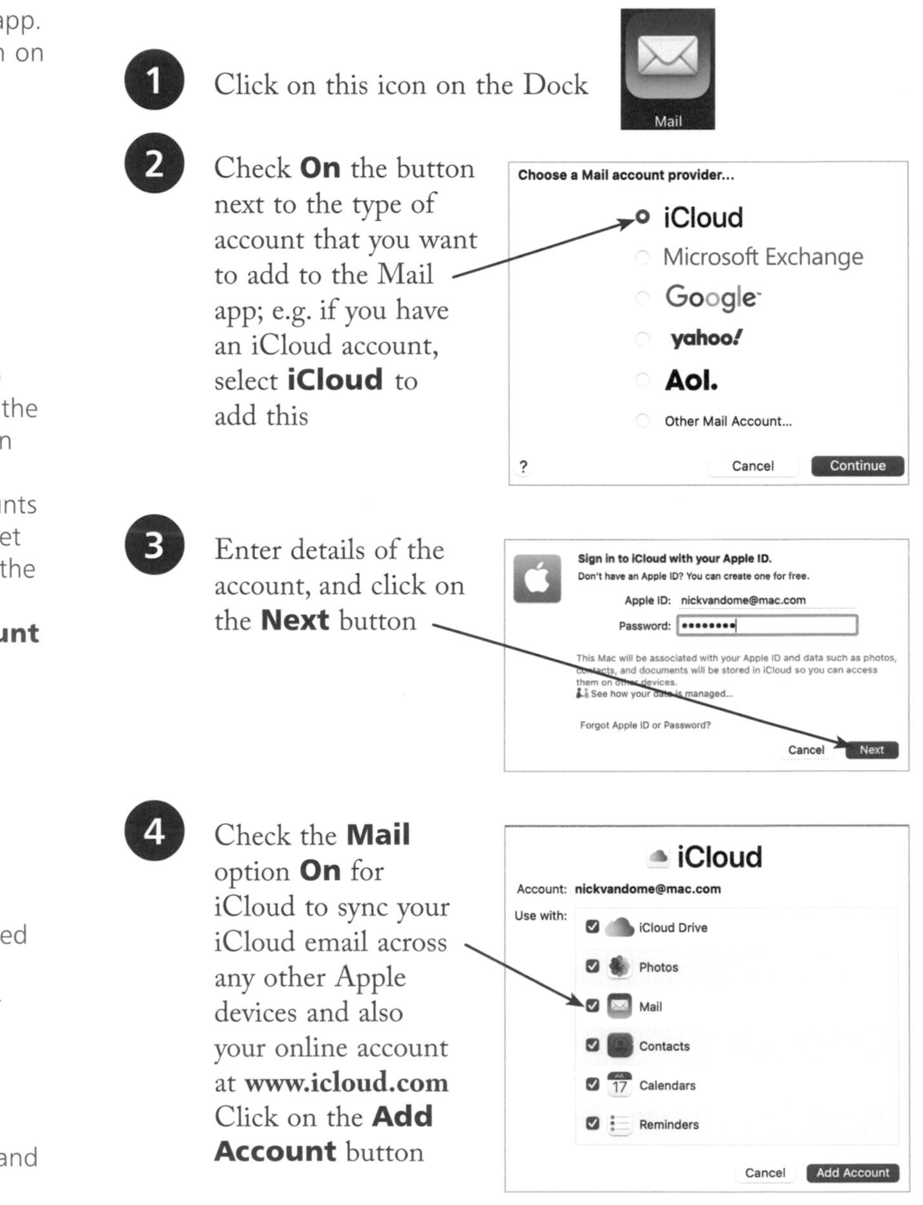

1. Click on this icon on the Dock

2. Check **On** the button next to the type of account that you want to add to the Mail app; e.g. if you have an iCloud account, select **iCloud** to add this

3. Enter details of the account, and click on the **Next** button

4. Check the **Mail** option **On** for iCloud to sync your iCloud email across any other Apple devices and also your online account at **www.icloud.com** Click on the **Add Account** button

Don't forget

You can set up more than one account in the Mail app and you can download messages from all of the accounts that you set up. To set up an account from the Mail app, select **Mail** > **Add Account** from the Menu bar.

Don't forget

Emails can be searched for using the Search icon in the top right-hand corner of the Mail app. Matching items are grouped according to People, Subjects, Mailboxes and Attachments.

Using Mail

Mail enables you to send and receive emails, and also format them to your own style. This can be simply formatting text, or attaching images or documents. To use Mail:

1. Click on the **Get Mail** button to download available email messages

2. Click on the **New Message** button to create a new email

3. Click on these buttons to **Reply** to, **Reply** (to) **All**, or **Forward** an email you have received

4. Select or open an email, and click on the **Trash** button to delete it

5. Click on the **Junk** button to mark an email as junk or spam. This trains Mail to identify junk mail. After a period of time, these types of messages will automatically be moved straight into the Junk mailbox

6. Click on the **Format** button to access options for formatting the text in an email

7. Use these buttons in the New Message window to select fonts and font size; color; bold; italic; underlining; strikethrough; and alignment options

8. Click on the **Attach** button to browse your folders, to include a file in your email. This can be items such as photos, Word documents or PDF files

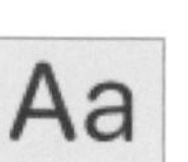

To show the descriptive text underneath an icon in Mail, **Ctrl** + **click** next to an icon and select **Icon and Text** from the menu.

When entering the name of a recipient for a message, Mail will display details of matching names from the Contacts app. For instance, if you type "DA", all of the entries in your Contacts app beginning with this will be displayed, and you can then select the required one.

If you forward an email with an attachment, then the attachment is included. If you reply to an email, the attachment will not be included.

Messaging

The Messages app enables you to send text messages (iMessages) to other macOS Monterey users or those with an iPhone, iPad or iPod Touch using iOS or iPadOS. It can also be used to send photos and videos. To use Messages:

1. Click on this icon on the Dock

2. Click on this button to start a new conversation

3. Click on this button and select a contact (this will be from your Contacts app).

To send an iMessage, the recipient must have an Apple ID

4. The person with whom you are having a conversation is displayed in the left-hand panel. The conversation continues down the right-hand panel

5. Click in the text box at the bottom of the main panel to write a message, and press **Return** on the keyboard to send

How's it going?

The Messages app has been updated in macOS Monterey, to include Shared with You (see page 137).

iMessages are specific to Apple and sent over Wi-Fi. SMS (Short Message Service) messages are usually sent between cellular devices, such as smartphones, that are contracted to a compatible mobile services provider. SMS messages can also be sent with the Messages app, to Apple or non-Apple users.

Audio messages can also be included in an iMessage. Click on this icon to the right of the text box, and record your message:

6. Click on this button at the right-hand side of the text box to access a range of emojis (small graphical symbols)

7. Click on an emoji to add it to a message

8. Scroll left and right of the window to view all of the emojis, or click on the bottom toolbar to move through the categories

9. Some emojis of people have options to select different styles. Click and hold on a relevant item to view the options. Click on one to add it to a message

Messages in macOS also supports Tapback, whereby you can add an icon to a message as a quick reply. To do this, click and hold on a message text and click on an icon.

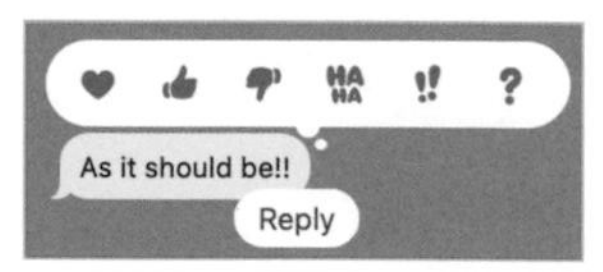

Pinning conversations

It is possible to keep your favorite conversations at the top of the left-hand panel by pinning them there. To do this:

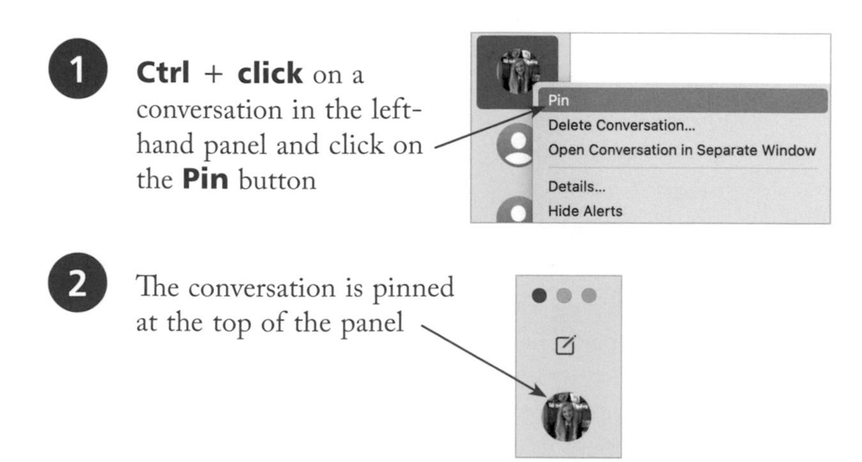

1. **Ctrl** + **click** on a conversation in the left-hand panel and click on the **Pin** button

2. The conversation is pinned at the top of the panel

Adding Memojis

Memojis are customized graphics that can be added to an iMessage in a similar way to an emoji. However, Memoji categories can also be created based on your own design, which can include a Memoji of yourself. To do this:

Photos and videos can also be added to messages, from the **Photos** button in Step 2.

1. Click on this icon next to the text box

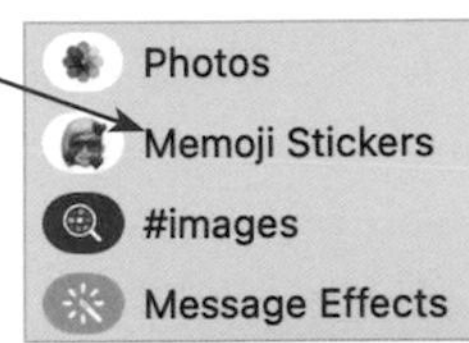

2. Click on the **Memoji Stickers** option

3. Click on the items on the top toolbar to view the different categories of Memoji. Click on a Memoji to add it to a message

The image in Step 5 is the basis for a Memoji, but it is not available on its own. Instead, variations are available; see Step 6.

4. Click on this button in the top left-hand corner of the Memoji window, to create a new, customized, Memoji category

5. A plain avatar for the Memoji is displayed in the main window. The left-hand sidebar contains categories for customizing the face. For each category, specific options appear below the face. Click on one to add it

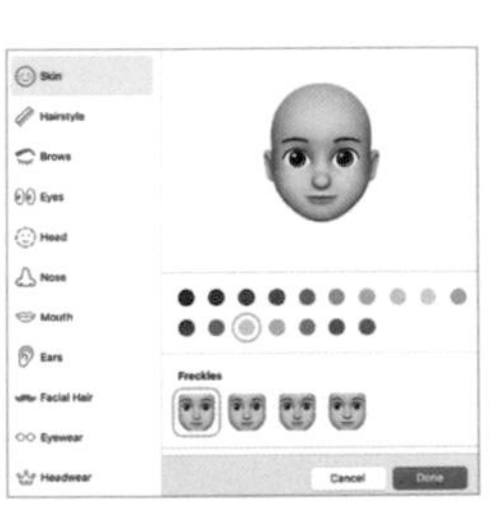

6. Add items to the avatar as required, using the categories in the left-hand sidebar. Click on the **Done** button to complete the Memoji, which is added to the Memoji Stickers panel

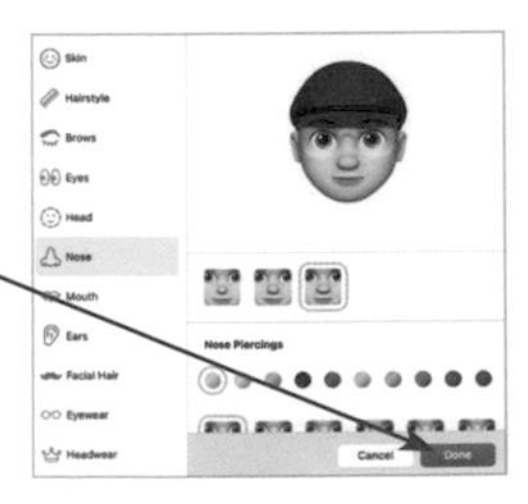

Shared with You

Text messages that you receive with the Messages app can contain a range of content other than text, including photos, audio clips, links to news items, links to websites, and music. When you receive this type of content, it can be downloaded to your Mac and you will be able to access it from the relevant app, using the Shared with You function; i.e. the Photos app for photos or videos, Safari for web links, and so on. The apps that support Shared with You are: Photos, Safari, News, Music, Podcasts and Apple TV. To use Shared with You:

Shared with You is a new feature in macOS Monterey.

1. Content in a text message in the Messages app can contain items such as photos or a web link

2. Or, it can contain a link to a story within the News app

apple.news

3. Items in the Messages app that are compatible with the Shared with You feature are accessed from the **Shared with You** section of the relevant app, which is usually accessed in the app's sidebar, such as within Safari, or the News app

3 Tabs
Received Links
Shared with You

Apple News
Today
News+
Shared with You

4. Click on an item within the Shared with You section to send a text message to the person who sent the item, without having to separately open the Messages app

If numerous photos are sent to you in the Messages app, they will all be available in the For You section of the Photos app. In addition, they can also be viewed in **Grid** view in the Messages app itself, by tapping once on this button next to the photos (this is a new feature in macOS Monterey):

Video Chatting with FaceTime

The FaceTime app has been comprehensively updated in macOS Monterey.

To make video calls with FaceTime you need an active internet connection and to be signed in with your Apple ID.

When you receive a FaceTime call, click on the **Accept** button to connect the call. Click on the **Decline** button to refuse it, or click on the down-pointing arrow next to the Decline button to select an option for replying to the invite or to set a reminder.

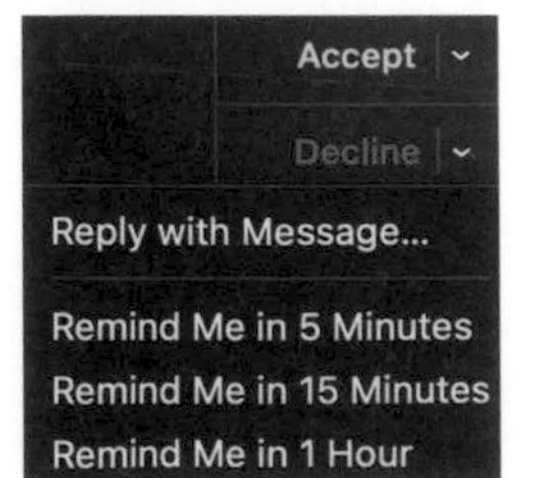

Video chatting is a very personal and interactive way to keep in touch with family and friends around the world. To use FaceTime for video chatting:

1. Click on the **FaceTime** app

2. Click on the **New FaceTime** button

3. Enter the name of a contact (or select one of the **Suggested** options)

4. Click on the **FaceTime** button to make a FaceTime call (to another FaceTime user: the **Create Link** in Step 2 can be used to make FaceTime calls to non-FaceTime users; see page 140)

5. When you have connected, your contact appears in the main window and you appear in a Picture in Picture thumbnail in the corner. Click and drag your own thumbnail to move it around the screen. The orientation of the FaceTime window depends on the one being used by the other person on the call

6. Use these buttons during a call to, from left to right: access the FaceTime sidebar (see next step); mute or unmute the microphone; turn the camera on or off for the call (audio will still be available if the camera is off); use SharePlay (see pages 141-142); and end the call

7. Click on the FaceTime sidebar button in Step 6 above to access more options for the call, including, from top to bottom: adding more people to the call; sharing links to the call for non-FaceTime users; and silencing requests to join FaceTime calls to which you are invited

The **Share Link** option in Step 7 can be used to create FaceTime calls with non-FaceTime users, including those using the Windows and Android operating systems. See page 140 for details.

Microphone modes

FaceTime also has options for how the microphone operates.

1. Start a FaceTime call, access the Control Center (see page 36) and click on the **Mic Mode** button

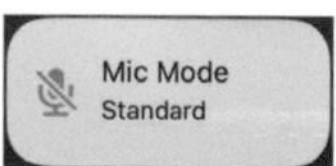

2. Click on the **Voice Isolation** button to block out background noise and give clearer prominence to the current speaker in a call

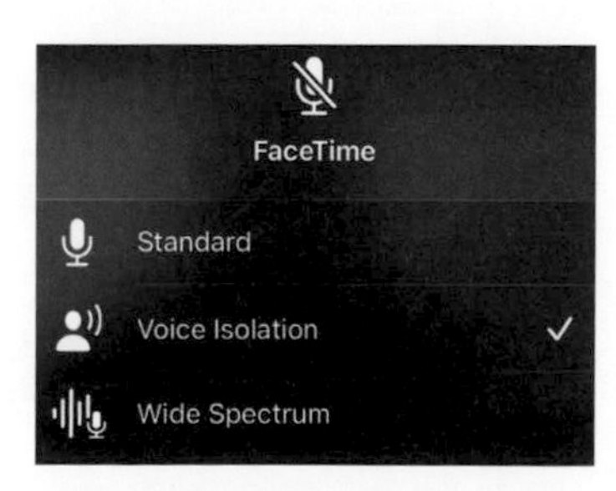

Spatial sound is used in group calls to present the sound as coming from the direction of the screen in which someone's FaceTime window is positioned, to create a more natural effect. This is a new feature in macOS Monterey.

Voice Isolation is a new feature in macOS Monterey.

Creating links for users without the FaceTime app to join a FaceTime call is a new feature in macOS Monterey.

It is also possible to invite someone to an existing FaceTime call, via a link, using the **Share Link** button in Step 7 on page 139.

When someone receives an invitation to a FaceTime call, the recipient can tap once on the **FaceTime Link** button in the invitation, enter their name and tap once on the **Continue** button.

Creating a FaceTime link

Before macOS Monterey, it was only possible to use FaceTime to make video calls to other FaceTime users, using an Apple device. However, that has now changed and the functionality of FaceTime has been expanded considerably, with the inclusion of an option to send a link to anyone, who can then join the FaceTime call via the web. To do this:

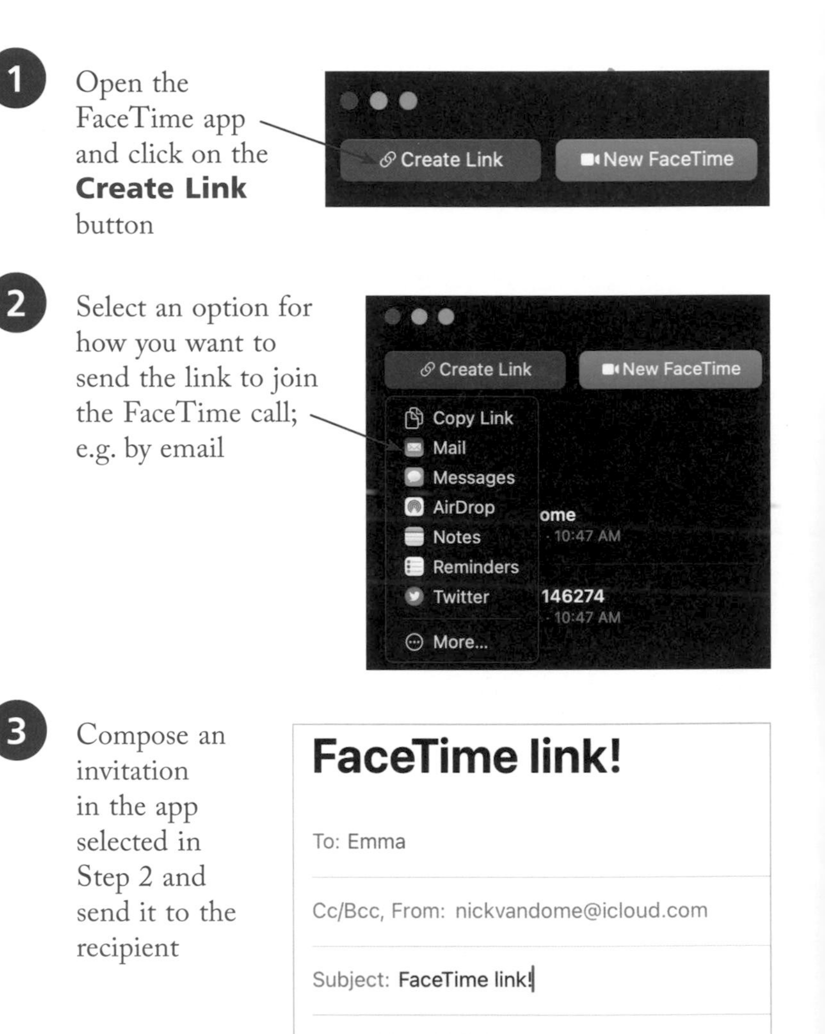

1. Open the FaceTime app and click on the **Create Link** button
2. Select an option for how you want to send the link to join the FaceTime call; e.g. by email
3. Compose an invitation in the app selected in Step 2 and send it to the recipient

Using SharePlay in FaceTime

SharePlay is a function within FaceTime that enables you to share your screen and to play movies and TV shows, or play music, and share this content with other people on a FaceTime call. To do this:

SharePlay is a new feature in macOS Monterey.

1. Within the FaceTime app, select **FaceTime** > **Preferences** from the top Menu bar. Click on the **SharePlay** tab and check **On** the **SharePlay** checkbox

2. Start a FaceTime call and click on this button on the FaceTime control panel to share your Mac screen with everyone else on an existing call

3. Click on the **Share My Screen** button

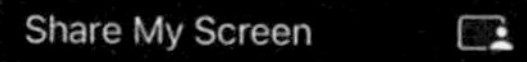

4. When a screen is being shared in a FaceTime call with SharePlay, whatever is shown on the sharing screen will be visible on everyone else's screen too. This appears as a separate panel, and the user's video feed is automatically turned off

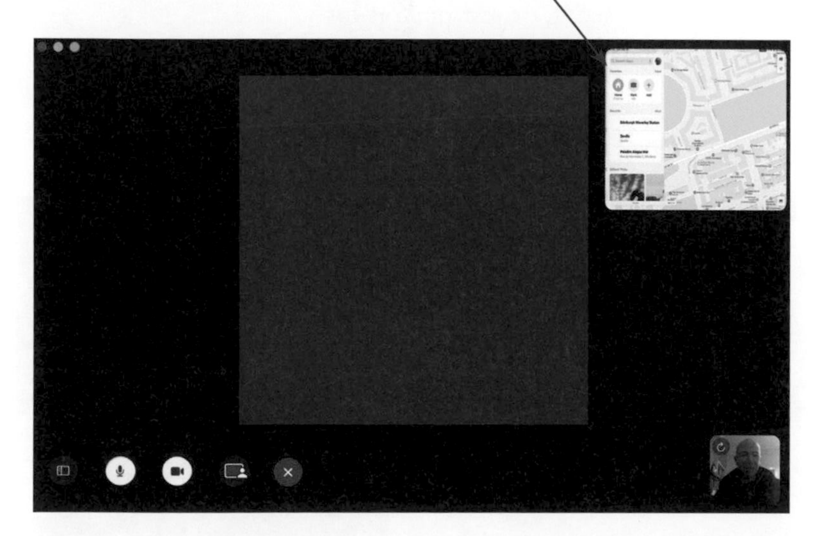

...cont'd

Sharing music, movies or TV shows

SharePlay can also be used to share content with other people on a FaceTime call from the Music app and the TV app. To do this:

Content for SharePlay is synchronized for everyone in a FaceTime call; so everyone sees, and hears, the same content at the same time.

1. Start a FaceTime call. and open either the **Music** app or the **TV** app. For either one, open the item you want to share on the FaceTime call

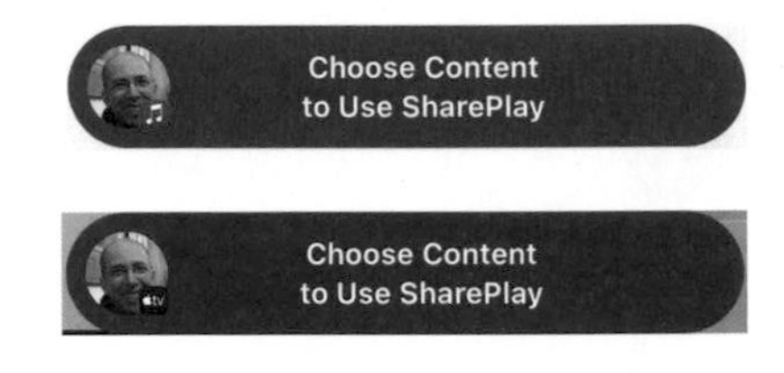

If content is being shared with SharePlay from the TV app, all participants can use the TV app controls; e.g. everyone can pause/play, rewind or fast forward, and this will be applied for everyone in the call.

2. For music, click on the SharePlay button and select how you want to start sharing music each time

3. For the **TV** app, select how you want to use an item you are opening; i.e. use it with SharePlay so that people on the FaceTime call can see it, or just open it for yourself on your Mac

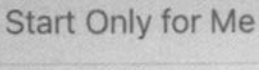

The controls at the top of the window in Step 4 can be used to make the content full-screen and adjust the volume. This only affects the device on which the changes are being made, not everyone in the call.

4. Content from the **TV** app that is used with SharePlay plays within the FaceTime interface and everyone on the call will be able to view it

8 Digital Lifestyle

Leisure time, and how we use it, is a significant issue for everyone. This chapter details some of the options with macOS, including photos, music, movies and reading books.

Using the Photos App

The Photos app is designed to mirror the one used on iOS devices and integrate more with iCloud, so that you can store all of your photos in iCloud and then view and manage them on all of your Apple devices.

If you are using the Photos app, you can specify how it operates with iCloud in its preferences.

1. Open the Photos app and click on **Photos** > **Preferences** on the top toolbar

2. Click on the **General** button to apply settings for items including how Memories are displayed (see page 146), how items are imported and whether videos and animated photos are played automatically

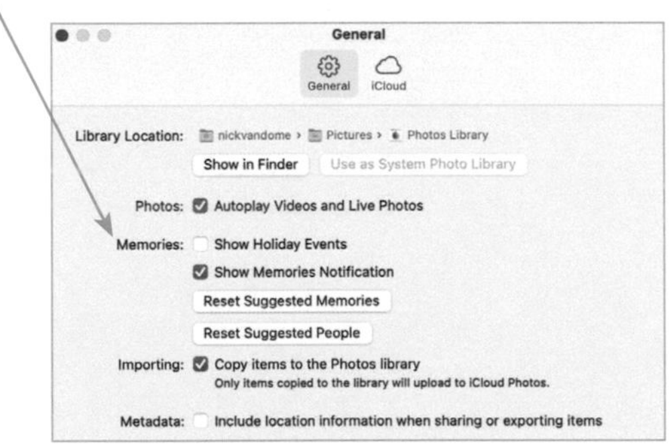

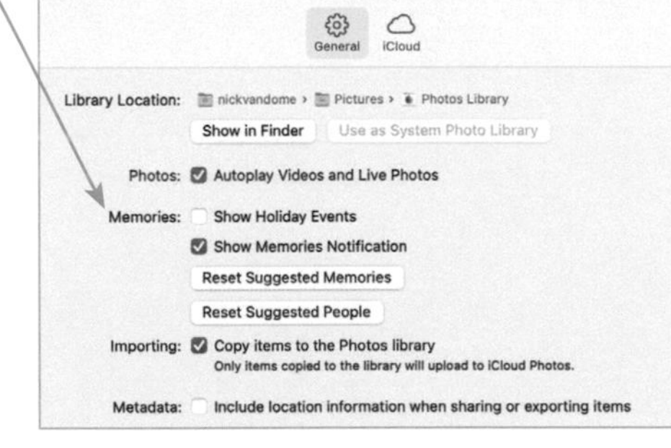

3. Click on the **iCloud** button for settings for using iCloud with the Photos app

The iCloud options for the Photos app are: **iCloud Photos**, which uploads your entire photo library to iCloud; **My Photo Stream**, which uploads the last 30 days' worth of photos so that they can be viewed on other Apple devices, using the same Apple ID; and **Shared Albums**, which enables you to share photos from the Photos app with family and friends.

Viewing Photos

The Photos app can be used to view photos according to years, months, days or at full size. This enables you to view your photos according to dates and times at which they were taken.

When you first open the Photos app you can select current photo libraries to import. This will ensure that all of the photos you want are available in the Photos app when you start.

1. Click on the **Library** button in the left-hand sidebar to see the various categories for viewing photos

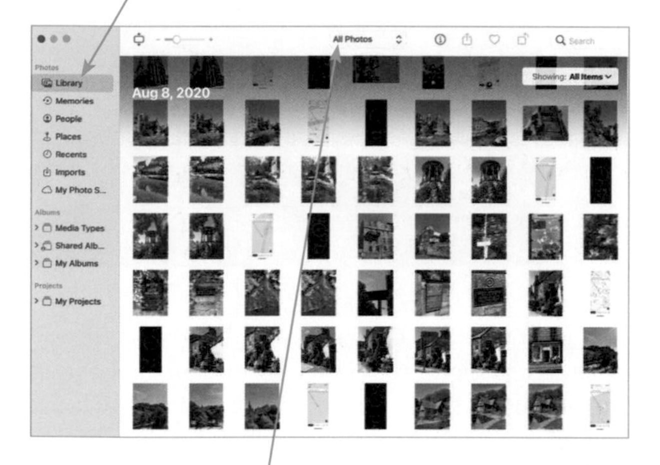

2. Click on the **All Photos** button on the top toolbar and select options for viewing photos

3. For each category – e.g. **Months** – the photos are displayed with the photos collated in easily-accessible formats

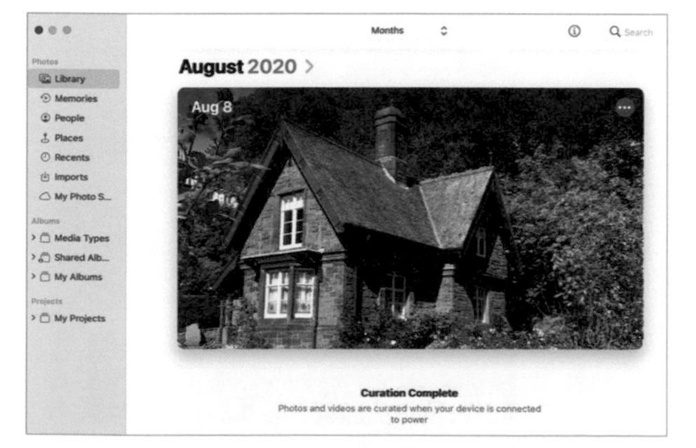

Photos can be imported into the Photos app by selecting **File** > **Import** from the Menu bar and navigating to the required location within the Finder. This can be used to import photos from your Mac or an external device connected with an USB cable, such as a digital camera, a card reader or a flashdrive.

...cont'd

Don't forget

Drag this slider to change the magnification of the photo(s) being displayed, in **Days** or **All Photos** view:

4. Double-click on a photo in the **All Photos**, **Days** or **Months** section to view it at full size

Hot tip

Click on the **Memories** button in the sidebar to view collections of related photos that have been collated by the Photos app, based on date and the best photos. Double-click on a Memory to view the photos within it. Click on this button in the bottom right-hand corner to view a Memory as a slideshow, with music and themes for transitions between each photo:

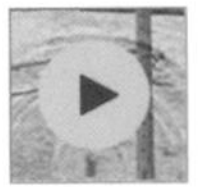

5. For a photo being displayed at full size use these buttons, from left to right, to: view information about it; share the photo; add it as a favorite; rotate it; apply auto-enhancement; and edit it

6. Click on the **Memories** button in the left-hand sidebar. This displays photos collated from specific dates (see the Hot tip)

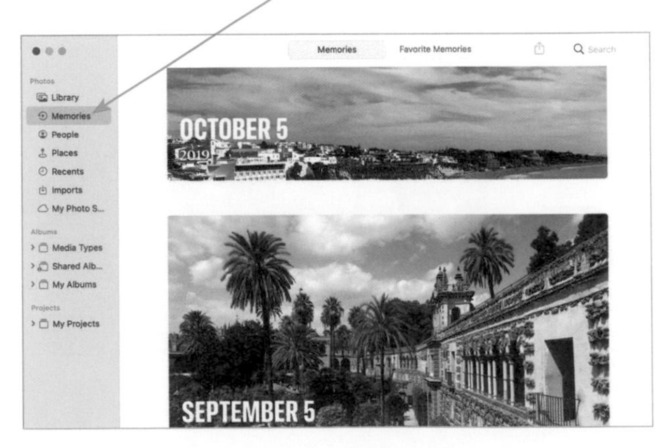

Using the sidebar

In the Photos app with macOS Monterey, the sidebar at the left-hand side is available regardless of the view being used; e.g. if you are looking at a Memory, or a photo at full size, the sidebar will still be available. To view categories and items in the sidebar:

1. Click on a category in the sidebar to view the items in the main window

2. The sidebar categories have their own toolbar at the top of the window. For instance, the **Places** category shows locations at which photos have been taken, and this can be viewed in Map, Satellite or Grid format

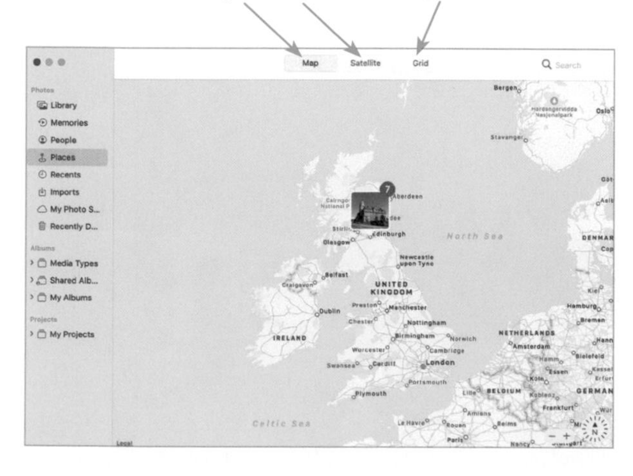

Click on the **People** or **Places** button in the sidebar to view collections of photos of specific people (based on the results of the Photos app scanning all of the available photos and identifying people within them) or locations where photos have been taken.

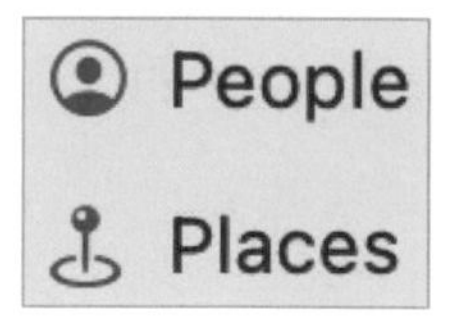

Editing Photos

The Photos app has a range of editing options that can be used to enhance your photos. To access these options:

1. Open a photo at full size
2. Click on the **Edit** button 

3. The main editing options are accessed from the **Adjust**, **Filters** and **Crop** buttons at the top of the window. Click on one of these to access its options in the right-hand panel

Click on the **Done** button at the top of the Photos app editing window to apply any editing changes that have been made.

4. For the **Adjust** options there is a slider that can be used to apply the amount of the required effect

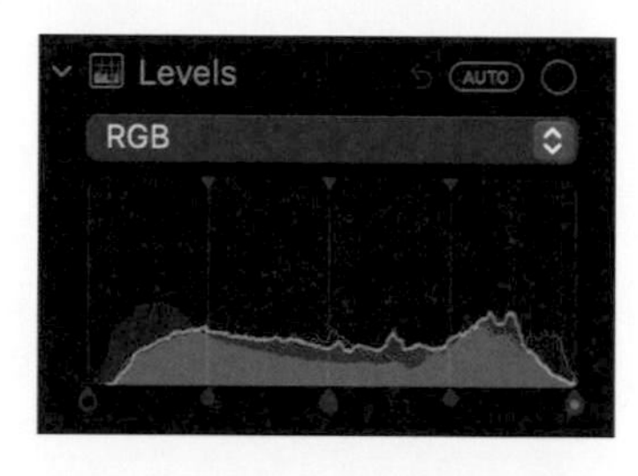

5. The **Adjust** options also have an **Options** section that can be used to fine-tune the editing effects, such as by dragging the markers at the bottom of the Levels graph

6. Click on the **Filters** button to access preset filters that can be applied to the photo. Click on a filter to view the effect in the photo

Hot tip

In the Photos app, Live Photos can also be edited. These are short, animated photos that can be captured on an iPhone or an iPad.

7. Click on the **Crop** button to crop the photo by dragging the corner markers. Drag on the dial to the right of the photo to rotate it

8. Click on the **Done** button to apply the effects

Starting with the Music App

Music is a major part of the Apple ecosystem, and for many years iTunes has been integral to this. However, in macOS Monterey, the iTunes app has been removed and replaced with a dedicated Music app. (Other forms of media content – e.g. movies and TV shows – also now have their own dedicated apps – e.g. the TV app or the Podcast app.) To use the Music app:

1. Click on the **Music** app on the Dock

2. Under the **Library** heading in the left-hand sidebar, click on a category to view the available songs; e.g. by **Artists**

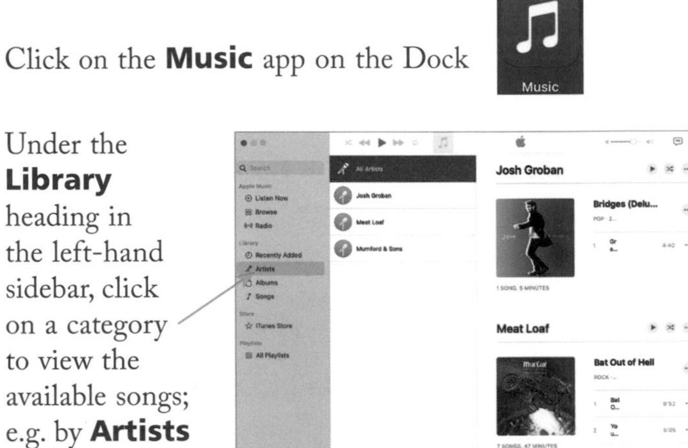

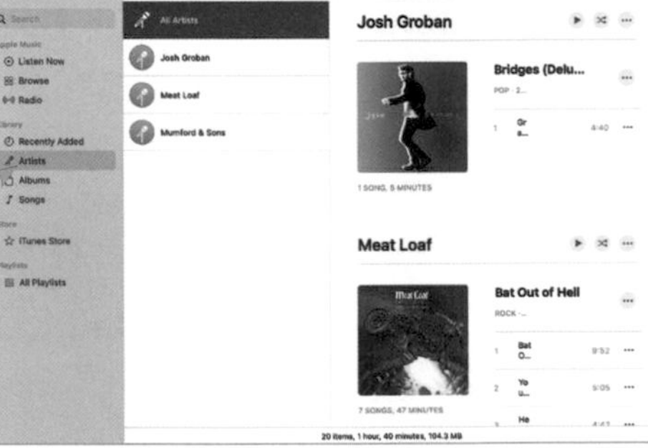

3. Click on a song or artist to view specific songs. Click on one to play it

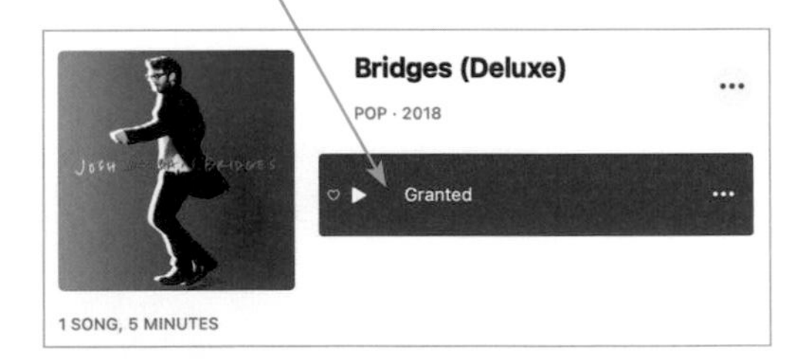

4. Use these controls to, from left to right: shuffle songs; play previous; play/pause; play next; and repeat

5. Drag this slider to change the volume

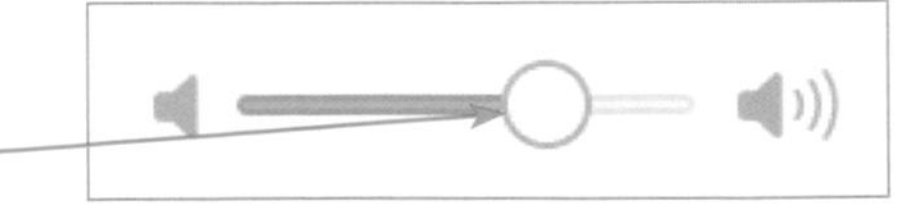

Never import music and use it for commercial purposes as this would be a breach of copyright.

Click on the **iTunes Store** button in the left-hand sidebar to access the online store for buying more music

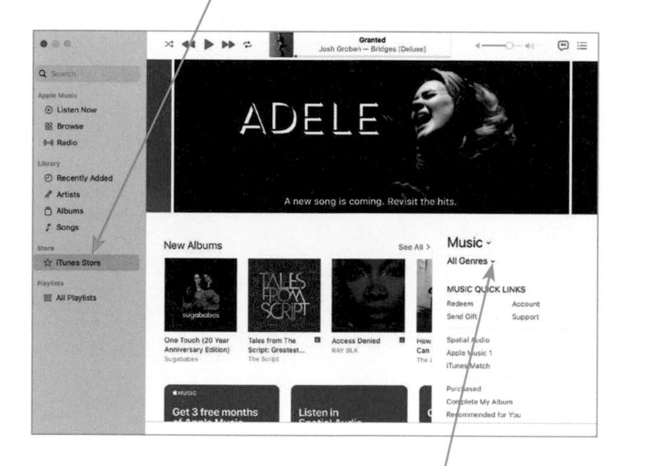

7. Navigate around the iTunes Store Homepage using the panels and sections within the Homepage. Click here to access different music categories

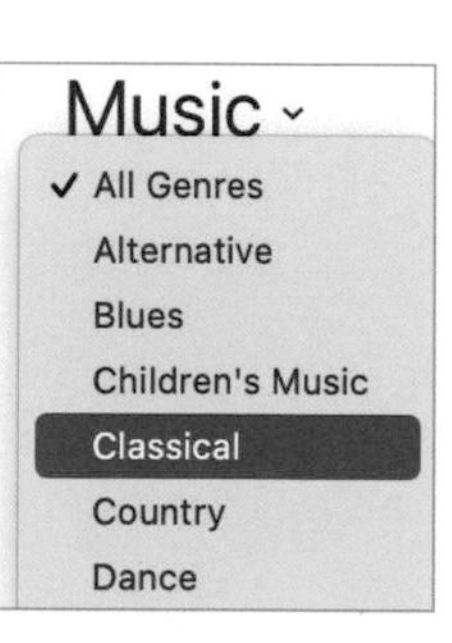

8. Click on a specific artist or album to view the available tracks. Click here to buy a whole album or click on individual songs to buy them

Music > Pop > The Script

Tales from The Script: Greatest Hits

The Script >

Songs | Ratings and Reviews | Related

£9.99 Buy

★★★★☆ (7)
Released 1 Oct, 2021
℗ 2021 Sony Music Entertainment UK Limited

	NAME	ARTIST	TIME	POPULARITY	PRICE
1.	Breakeven	The S...	4:20		£0.99
2.	The Man Wh...	The S...	4:00		£0.99
3.	For the Fir...	The S...	4:12		£0.99
4.	Nothing	The S...	4:31		£0.99
5.	Hall of Fame...	The S...	3:22		£0.99
6.	If You Coul...	The S...	3:38		£0.99
7.	Superheroes	The S...	4:02		£0.99
8.	Six Degree...	The S...	3:52		£0.99
9.	Rain	The S...	3:30		£0.99
10.	Arms Open	The S...	4:02		£0.99

Don't forget

Although iTunes is no longer a separate app in macOS Monterey, the iTunes Store is still available from the Music app. This now just contains music, rather than the variety of content that was previously available in the iTunes Store. Once items have been bought, they are available in the Music app, as shown in Step 2 on the previous page.

Never use illegal music or video download sites. Apart from the legal factor, they are much more likely to contain viruses and spyware.

Using Apple Music

Apple Music is a service that makes the entire Apple Music library of music available to users. It is a subscription service, and music can be streamed over the internet or downloaded so that you can listen to it when you are offline. To start with Apple Music:

Numerous radio stations can also be listened to with Apple Music, including Beats 1.

1. Click on the **Music** app on the Dock

2. Click on the **Listen Now** button underneath the Apple Music heading in the left-hand sidebar

To cancel your Apple Music subscription at any point, open the App Store app and click on your account icon at the bottom of the left-hand sidebar. Click on the **View Information** button at the top of the main window and log in with your Apple ID details. Scroll to the **Manage** section and click on the **Manage** button to access details of your Apple subscriptions. Click on the **Edit** button next to the subscription that you want to cancel and click on the **Cancel Subscription** button.

3. Click on the **Try It Free** button to select a free three-month trial of the service

4. Once the free trial has ended, click on one of the subscription options

Apple TV

Apple TV is an online service for downloading movies and TV shows. These were previously accessed from the iTunes app. Content can be bought or rented, depending on the item, and there is also access to the subscription TV service, Apple TV+. The content from Apple TV is accessed from the TV app.

Apple TV+ can also be accessed from the TV app. This is a monthly subscription service that provides exclusive movies and TV shows, created by Apple.

Click on the **TV** app on the Dock

2 The initial content is displayed in the **Watch Now** window, with recommended content

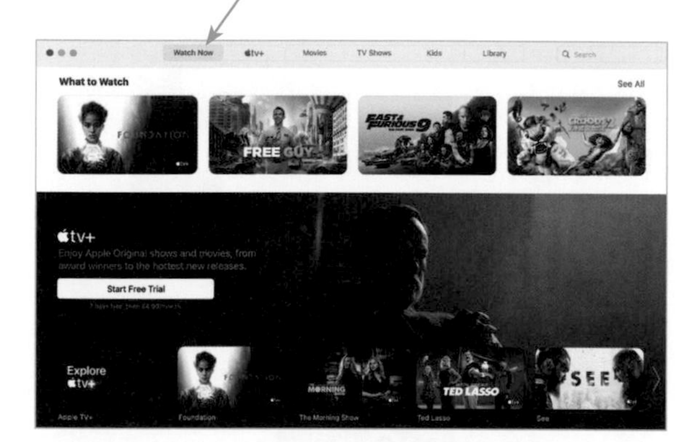

Use these buttons on the top toolbar to access more content, including **TV+**, **Movies**, **TV Shows** and **Kids**

Watch Now | tv+ | Movies | TV Shows | Kids | Library

Click on the **Library** button on the top toolbar to view movies and TV shows that you have downloaded with the TV app.

4 Movies and TV shows can be accessed from their relevant categories, and they can be rented or bought. TV+ is a subscription service, with a free trial. Click on the **Start Free Trial** button to start a 7-day free trial

If you rent a movie from the Apple TV store, you have 30 days to watch it and 48 hours to complete it once you have started watching it.

Podcasts

Podcasts are becoming increasingly widespread and are a popular way to consume a wide range of audio content, covering topics including news, sport and comedy. macOS Monterey uses the Podcasts app to provide access to a vast library of podcasts.

1. Click on the **Podcasts** app on the Dock

2. Click on the **Browse** button in the left-hand sidebar and swipe up and down to view the available podcasts

3. Click on a podcast to view its details. Click on the **+ Subscribe** button to subscribe to the podcast so that new episodes will be downloaded automatically

4. Click on these buttons in the left-hand sidebar to access more items in the Podcasts app

5. Click on these buttons in the left-hand sidebar to view podcasts that you have subscribed to, including the overall show and individual episodes

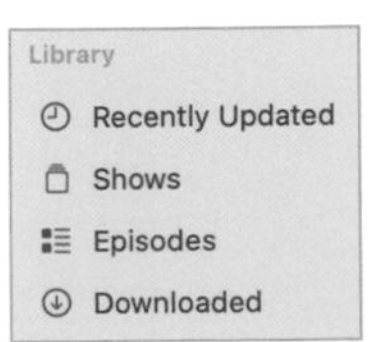

Don't forget

Most podcasts consist of regular episodes, usually weekly. Once you have subscribed to a podcast, each episode will be downloaded as it is released. To play a podcast, click on the **Shows** button in the left-hand sidebar and click on the **Play Episode** button.

Reading with Books

Books is an eBook reading app that has been available with Apple's mobile devices, including the iPhone and the iPad, for a number of years. macOS Monterey also offers this technology for desktop and laptop Macs. To use Books:

1. Click on the **Books** app on the Dock and sign in with your Apple ID

2. All of the currently-available books are displayed in the **Library**

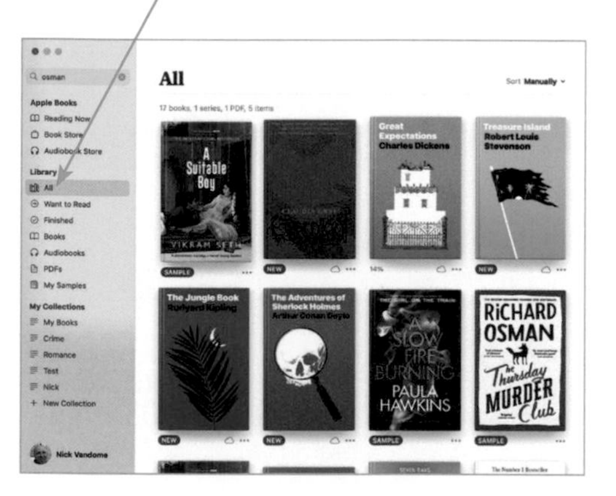

3. Click on the **Browse Sections** heading to view and download more books

Browse Sections

4. Click on a title to preview details about it

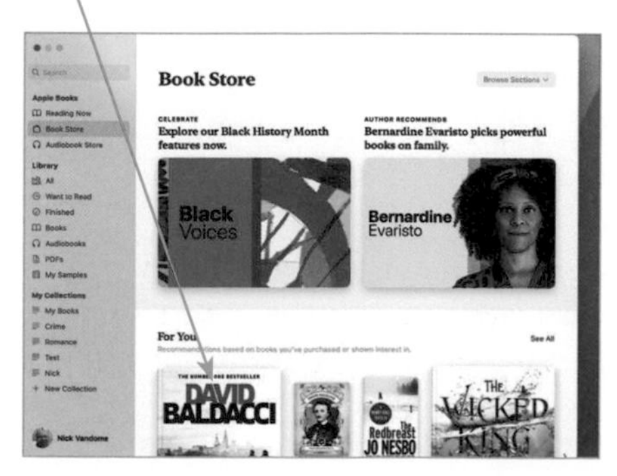

Don't forget

Books consists of your own library for storing and reading eBooks and also access to the online **Book Store** for buying and downloading new books. Items that you have downloaded with Books on other devices will also be available in your Books Library on your Mac.

...cont'd

5. Click here to download the book (if it is a paid-for title this button will display a price)

Use these buttons on the top toolbar of a book to, from left to right: view the Table of Contents; view any bookmarks that have been added; or view any notes that you have added.

6. The title will be downloaded into your Books Library. Double-click on the cover to open the book and start reading

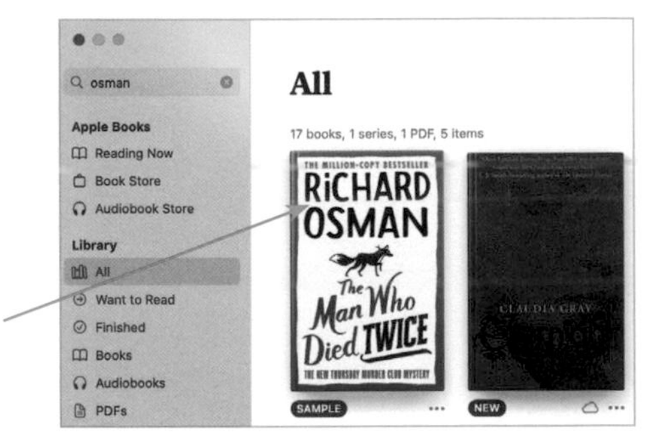

Use these buttons on the top toolbar to, from left to right: change the text size; search for text; and add a bookmark to a page.

7. Click or tap on the right-hand and left-hand edges to turn a page. Move the cursor over the top of the page to access the top toolbar. The bottom toolbar displays the page numbers and location

9 Sharing macOS

This chapter looks at how to set up and manage different user accounts on your Mac.

Adding Users

macOS enables multiple users to access individual accounts on the same computer. If there are multiple users (i.e. two or more for a single machine), each person can log in individually and access their own files and folders. This means that each person can log in to their own settings and preferences. All user accounts can be password-protected, to ensure that each user's environment is secure. To set up multiple user accounts:

Don't forget

Every computer with multiple users has at least one main user, also known as the administrator. This means that they have greater control over the number of items that they can edit and alter. If there is only one user on a computer, they automatically take on the role of administrator. Administrators have a particularly important role to play when computers are networked together. Each computer can potentially have several administrators.

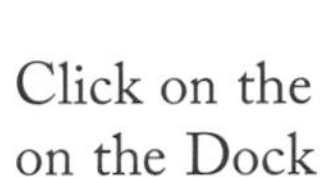

1. Click on the **System Preferences** icon on the Dock

2. Click on the **Users & Groups** icon

3. The information about the current user account is displayed. This is your own account, and the information is based on details you provided when you first set up your Mac

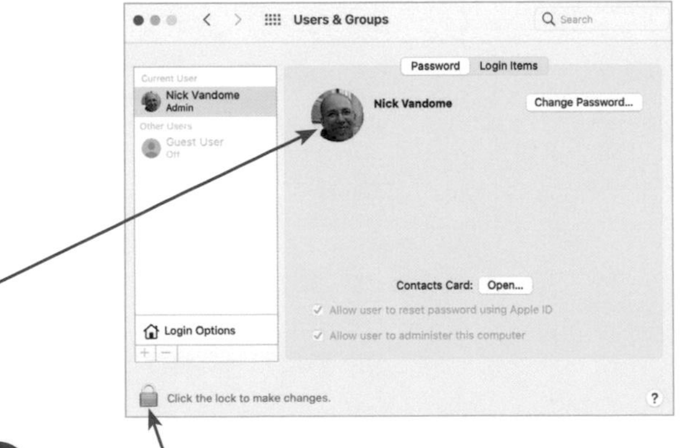

Each user can edit their own icon or photo of themselves, by moving the cursor over the current image and clicking on the **edit** button.

4. Click on this icon to enter your password to unlock the settings so that new accounts can be added

Click the lock to prevent further changes.

5. Click on the **+** button to add a new account

6. Enter the details for the new account holder

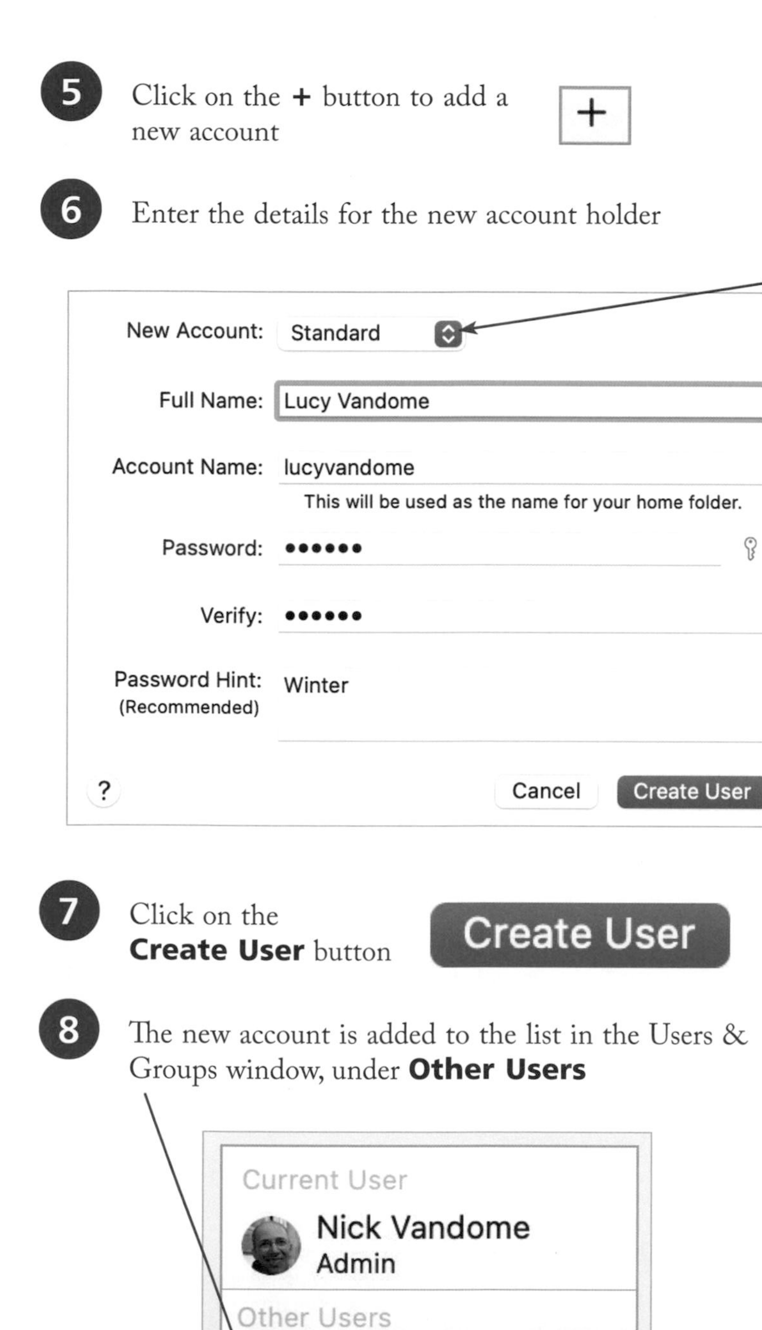

7. Click on the **Create User** button

8. The new account is added to the list in the Users & Groups window, under **Other Users**

At Step 6, choose the type of account from the drop-down list. An **Administrator** account is one that allows the user to make system changes, and add or delete other users. A **Standard** account allows the user to use the functionality of the Mac, but not change system settings. **Sharing Only** is an account that allows guests to log in temporarily, without a password – when they log out, all files and information will be deleted from the guest account.

By default, you are the administrator of your own Mac. This means that you can administer other user accounts.

Deleting Users

Once a user has been added, their name appears on the list in the Users & Groups window (see Step 8 on page 159). It is then possible to edit the details of a particular user or delete them altogether. To do this:

Always tell other users if you are planning to delete them from the system. Don't just remove them and then let them find out the next time they try to log in. If you delete a user, their personal files are left untouched and can still be accessed, if you select **Don't change the home folder** in Step 3.

1. Within **Users & Groups**, unlock the settings as shown in Step 4 on page 158, then select a user from the list

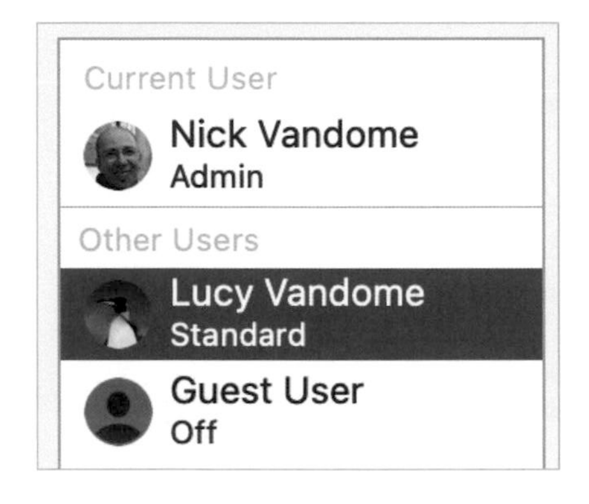

2. Click here to remove the selected person's user account

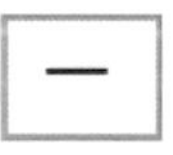

3. A warning box appears, to check if you really do want to delete the selected user. If you do, select the required option and click on the **Delete User** button

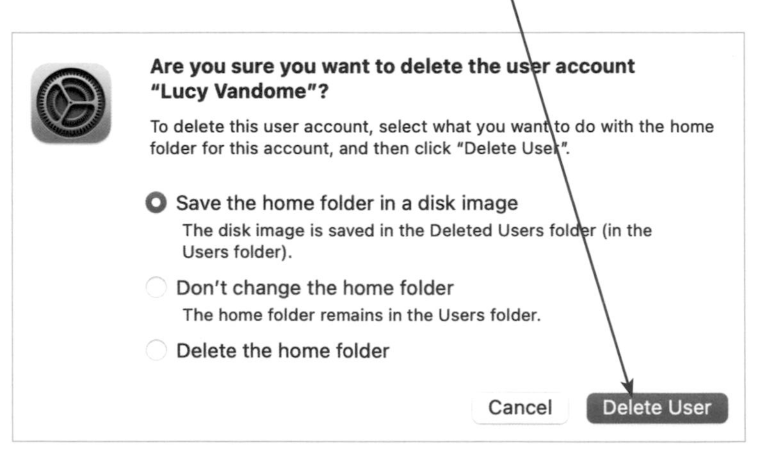

Fast User Switching

If there are multiple users on your macOS system, it is useful to be able to switch between them as quickly as possible. When this is done, the first user's session is retained so that they can return to it if required. To switch between users:

Unlock the settings before you start (see Step 4 on page 158).

1. In the Users & Groups window, click on the **Login Options** button

Login Options

2. Check **On** the **Show fast user switching menu as** box, then close the window

When you switch between users, the first user remains logged in, and their current session remains intact.

3. At the top right of the screen, click on the current user's name

Nick Vandome

4. Click on the name of another user

Nick Vandome

Lucy Vandome

Nick Vandome

Login Window...

Users & Groups Preferences...

Users can sign in from the Lock screen with a password by clicking on their own icon/name on the screen and entering the relevant details. If fast user switching is not used, each user has to log out before the next one can sign in.

5. Enter the relevant password (if required)

6. Click on this button to log in

Screen Time

The amount of time that we spend on our digital devices is a growing issue in society, and steps are being taken to let us see exactly how much time we are spending looking at our computer screens. In macOS Monterey, a range of screen-use options can be monitored with the Screen Time feature. To use this:

1. Select **System Preferences** > **Screen Time**

2. Click on the **Options** button

3. Click on the **Turn On...** button to activate Screen Time

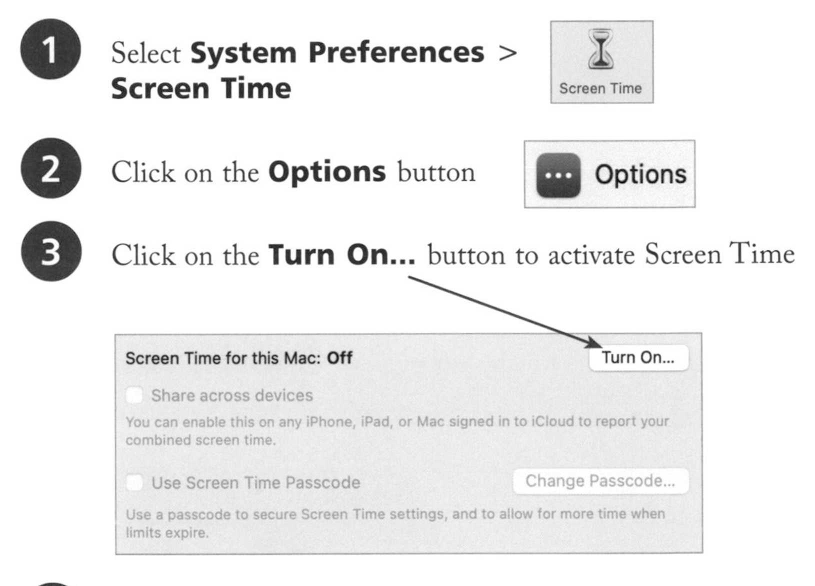

4. Click on the **App Usage** button in the left-hand sidebar to view the overall screen usage. Click here to select a different time period to view

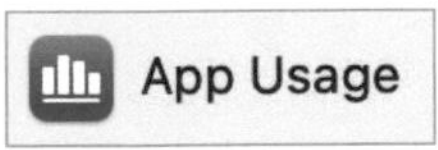

All of the Screen Time options can be accessed from the left-hand sidebar of the main window.

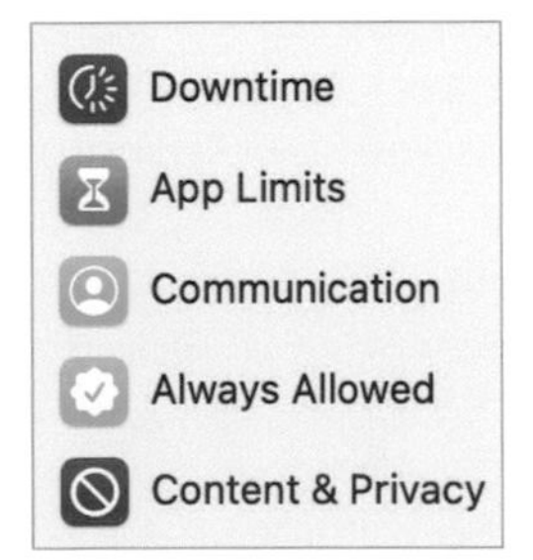

Downtime

This can be used to specify times when the computer and its apps cannot be accessed.

1. Check **On** the **Every Day** button to apply time limits for a whole week

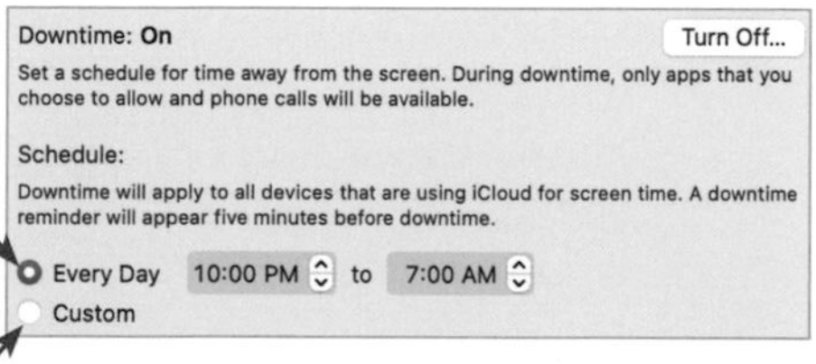

2. Check **On** the **Custom** button to apply time limits for individual days

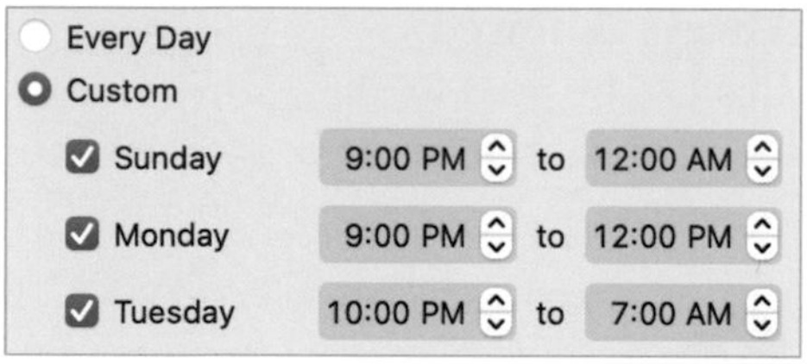

App Limits

This can be used to limit the amount of time that certain categories of apps can be accessed.

1. Click on the **+** button at the bottom of the App Limits window

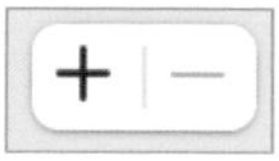

2. Click on any category to which you want limits to apply

3. Click here to set a time limit for the categories selected in Step 2. Click on the **Custom** option to apply custom times as for Downtime

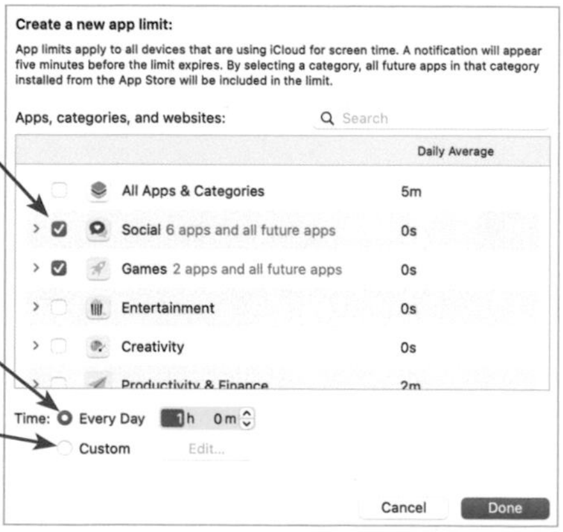

In general, the categories for app limits relate to the categories under which the apps are listed in the App Store.

...cont'd

Communication

This can be used to specify who can contact you during screen time.

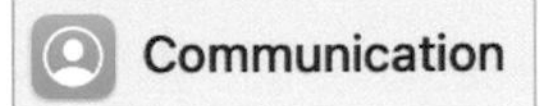

Click here to select who can contact you during screen time and also downtime, from contacts or everyone

Always Allowed

This can be used to allow the use of specified apps, regardless of other settings.

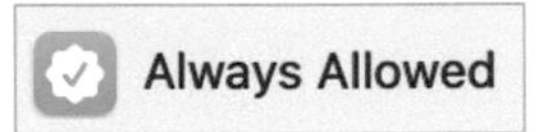

Check **On** the items that you want to allow, regardless of any other Screen Time settings

Content & Privacy

This can be used to add restrictions based on the type of content being viewed.

Click on the **Content** tab to restrict content based on criteria such as adult content and explicit language

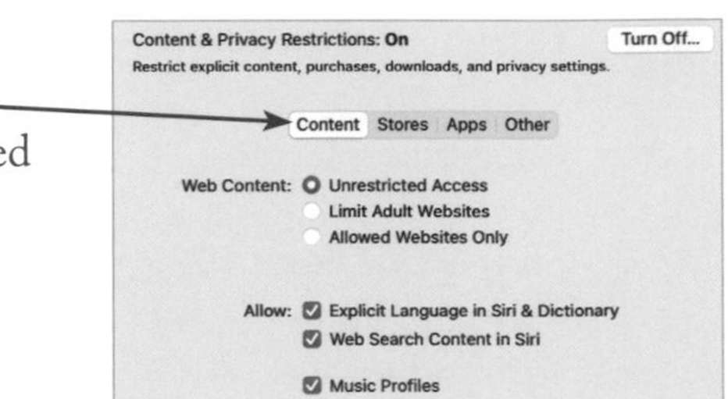

Click on the **Stores** tab under **Content & Privacy** to access options for content that is downloaded from the App Store, the Music app, the TV app and the Podcasts app. Selections can be made for age-appropriate content and also blocking content with explicit language. Click on the **Apps** button to allow the use of specific apps regardless of other privacy restrictions.

10 Networking

This chapter looks at networking and how to share files over a network.

Networking Overview

Before you start sharing files directly between computers, you have to connect them together. This is known as networking and can be done with two computers in the same room, or with thousands of computers in a major corporation. If you are setting up your own small network, it will be known in the computing world as a Local Area Network (LAN). When setting up a network there are various pieces of hardware that are initially required to join all of the network items together. Once this has been done, software settings can be applied for the networked items. Some of the items of hardware that may be required include:

Connecting to the internet is also another form of network connection.

- **A network card**. This is known as a Network Interface Card (NIC), and all recent Macs have them built in.
- **A wireless router**. This is for a wireless network, which is increasingly the most common way to create a network via Wi-Fi. The router is connected to a telephone line and the computer then communicates with it wirelessly.
- **An Ethernet port and Ethernet cable**. This enables you to make the physical connection between devices. Ethernet cables come in a variety of forms, but the one you should be looking for is the Cat6 type as this allows for the fastest transfer of data. If you are creating a wireless network then you will not require these.
- **A hub**. This is a piece of hardware with multiple Ethernet ports that enables you to connect all of your devices together and lets them communicate with each other. However, conflicts can occur with hubs if two devices try to send data through one at the same time.
- **A switch**. This is similar in operation to a hub but it is more sophisticated in its method of data transfer, thus allowing all of the machines on a network to communicate simultaneously, unlike a hub.

The latest models of MacBook do not have a separate Ethernet port. However, a Thunderbolt to Ethernet adapter can be used to connect an Ethernet cable to a MacBook.

Once you have worked out all of the devices that you want to include on your network, you can arrange them accordingly. Try to keep the switches and hub within relative proximity of a power supply, and if you are using cables, make sure they are laid out safely.

Wireless network

The most common method of creating a wireless network is by using Wi-Fi. This involves using a router to connect your Mac computer to the Wi-Fi connection, provided by an external supplier. Once the Wi-Fi connection has been made, the Mac will be able to join compatible online networks: the most common one being the internet. It can also be used to link other computers on a network. The Wi-Fi settings are available within System Preferences, in the Network category. See page 168 for details about connecting your Mac to a Wi-Fi network.

Status: **Connected** Turn Wi-Fi Off
Wi-Fi is connected to PLUSNET-TXJ5 and has the IP address 192.168.1.75.

Network Name: PLUSNET-TXJ5

- [x] Automatically join this network
- [x] Ask to join Personal Hotspots
- [ ] Ask to join new networks

Known networks will be joined automatically. If no known networks are available, you will have to manually select a network.

Another method for connecting items wirelessly is Bluetooth. This covers short distances and is usually used for items such as printers and smartphones. Bluetooth devices can be connected using the Bluetooth File Exchange utility: select the required files you want to exchange via Bluetooth and then select the required device.

Ethernet network

Another option for networking computers is to create an Ethernet network. This involves buying an Ethernet hub or switch, which enables you to connect several devices to a central point; i.e. the hub or switch. Older Apple computers and most modern printers have an Ethernet connection, so it is possible to connect various devices, not just computers. Once all of the devices have been connected by Ethernet cables, you can then start applying network settings. If a Mac computer does not have an Ethernet connection, an adapter can be used to connect an Ethernet cable to the Mac.

AirPort network

Apple previously used its own version of Wi-Fi routers to connect to Wi-Fi networks. This was known as AirPort, but was discontinued in 2018. However, these devices can still be used to connect to a Wi-Fi networks, although they are no longer supported by Apple.

Network Settings

Once you have connected the hardware needed for a network, you can start applying the network settings that are required for connecting to the internet, for online access.

1. In **System Preferences**, click on the **Network** button

2. For a wireless connection, click on the **Turn Wi-Fi On** button

3. Details of wireless settings are displayed

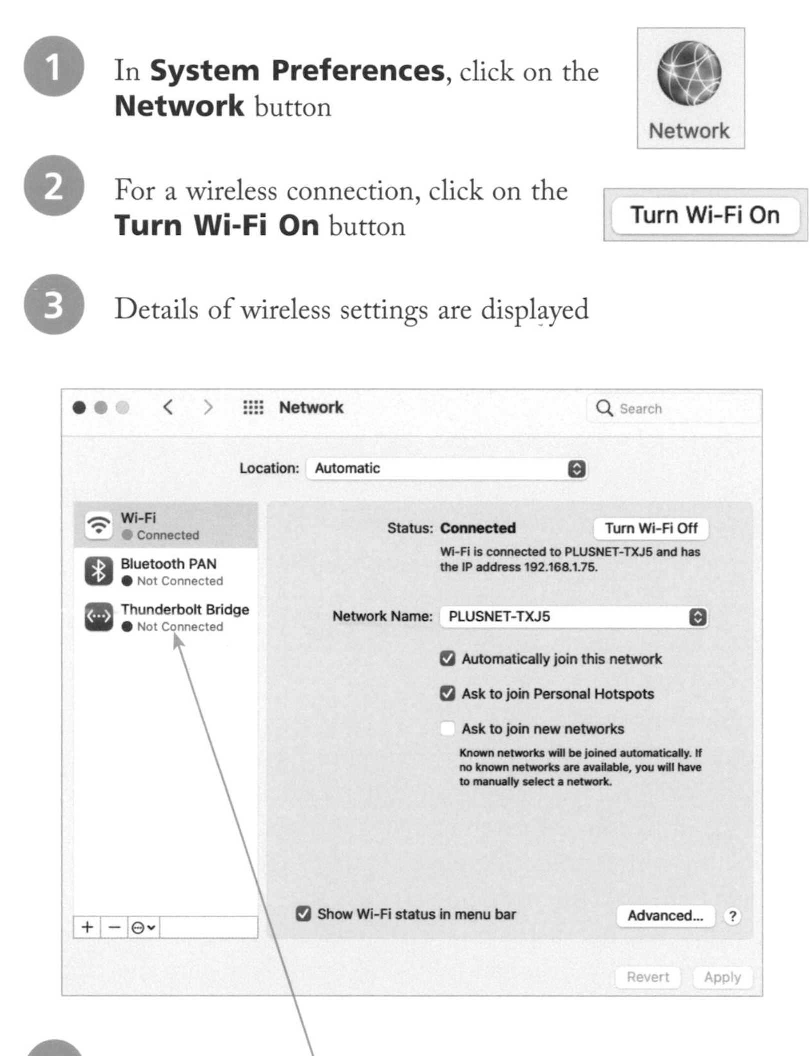

4. For a cable connection, connect an Ethernet cable (via an adapter in the Thunderbolt port) and click on the **Thunderbolt Bridge** option

5. Click on the **Advanced...** button to see the full settings for each option

An Ethernet cable can be used to connect to a router, using an adapter (see tip on page 166), instead of using a Wi-Fi connection, and it can also be used to connect two computers.

File Sharing

One of the main reasons for creating a network of two or more computers is to share files between them. On networked Macs, this involves setting them up so that they can share files, and then accessing these files.

Setting up file sharing

To set up file sharing on a networked Mac:

1. Click on the **System Preferences** button on the Dock

2. Click on the **Sharing** icon

3. Check **On** the boxes next to the items you want to share (the most common items to share are files and printers)

On	Service
☐	Screen Sharing
☑	File Sharing
☐	Media Sharing
☑	Printer Sharing
☐	Remote Login
☐	Remote Management
☐	Remote Apple Events
☑	Bluetooth Sharing
☐	Internet Sharing
☐	Content Caching

For macOS users, files can also be shared with the AirDrop option. This can be used with two Macs, and compatible iPhones and iPads that have this facility. Files can be shared simply by dragging them onto the icon of the other user that appears in the AirDrop window. To access AirDrop, click on this button in the Finder:

Connecting to a Network

Connecting as a registered user

To connect as a registered user (usually as yourself when you want to access items on another one of your own computers):

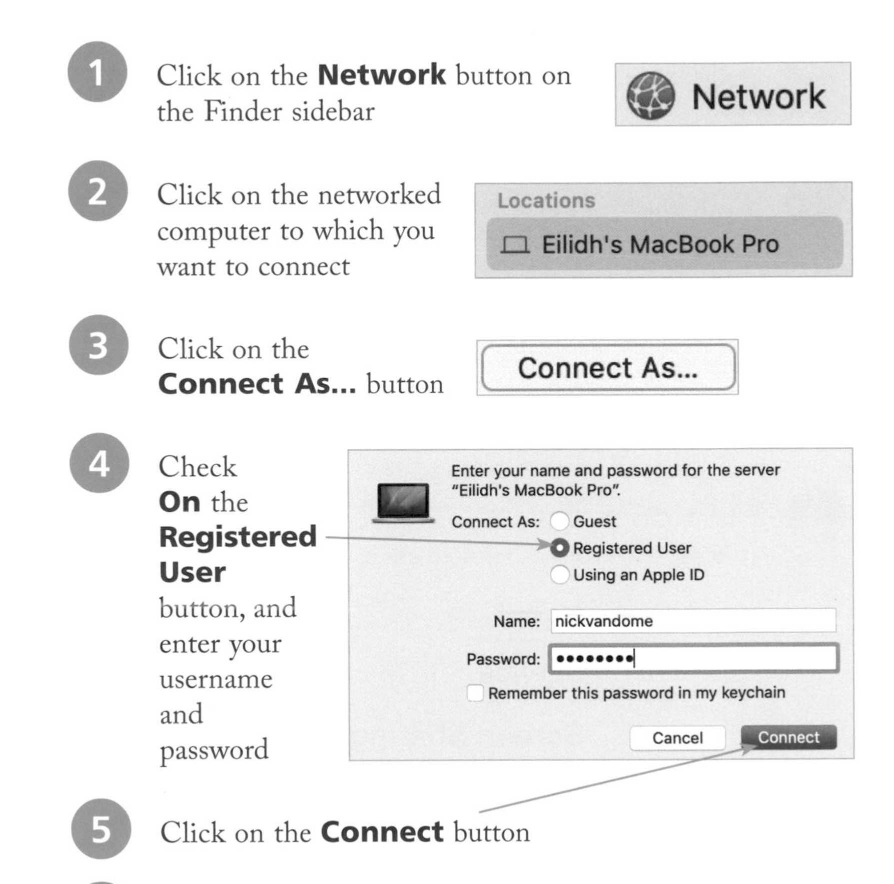

1. Click on the **Network** button on the Finder sidebar
2. Click on the networked computer to which you want to connect
3. Click on the **Connect As...** button
4. Check **On** the **Registered User** button, and enter your username and password
5. Click on the **Connect** button
6. The public folders and home folder of the networked computer are available to the registered user. Double-click on an item to view its contents

The username and password used to connect to a networked computer as a registered user are the ones used to log in to your Mac. There is also an option to connect using your Apple ID.

You can disconnect from a networked computer by ejecting it in the Finder in the same way as you would for a removable drive, such as a flashdrive (see page 53), by clicking on the **Eject** icon.

Locations
Nick's MacBook Pro
Remote Disc
Eilidh's MacBook...

Guest users

Guest users on a network are users other than yourself or other registered users, to whom you want to limit access to your files and folders. Guests only have access to a folder called the Drop Box in your own Public folder. To share files with guest users you have to first copy them into the Drop Box. To do this:

1. Create a file and select **File** > **Save** from the Menu bar

2. Navigate to your own home folder (this is created automatically by macOS and displayed in the Finder sidebar with your Mac username)

3. Double-click on the **Public** folder

4. Double-click on the **Drop Box** folder

5. Save the file into the Drop Box

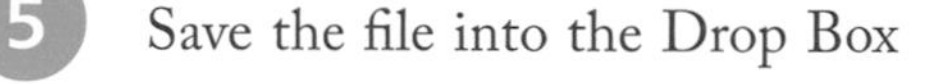

If another user is having problems accessing the files in your Drop Box, check the permissions settings that have been assigned to the files. See page 186 for further details.

The contents of the Drop Box can be accessed by other users on the same computer as well as users on the network.

The Drop Box folder is not the same as the Dropbox app, which is used for online storage and backing up.

...cont'd

Accessing a Drop Box

To access files in a Drop Box:

It is better to copy files into the Drop Box rather than moving them completely from their primary location.

1. Access a networked computer as shown on page 170
2. Click on the **Connect As...** button in the Finder window
3. Check **On** the **Guest** button
4. Click on the **Connect** button
5. Double-click on a user's **Public Folder**
6. Double-click on the **Drop Box** folder to access the files within it

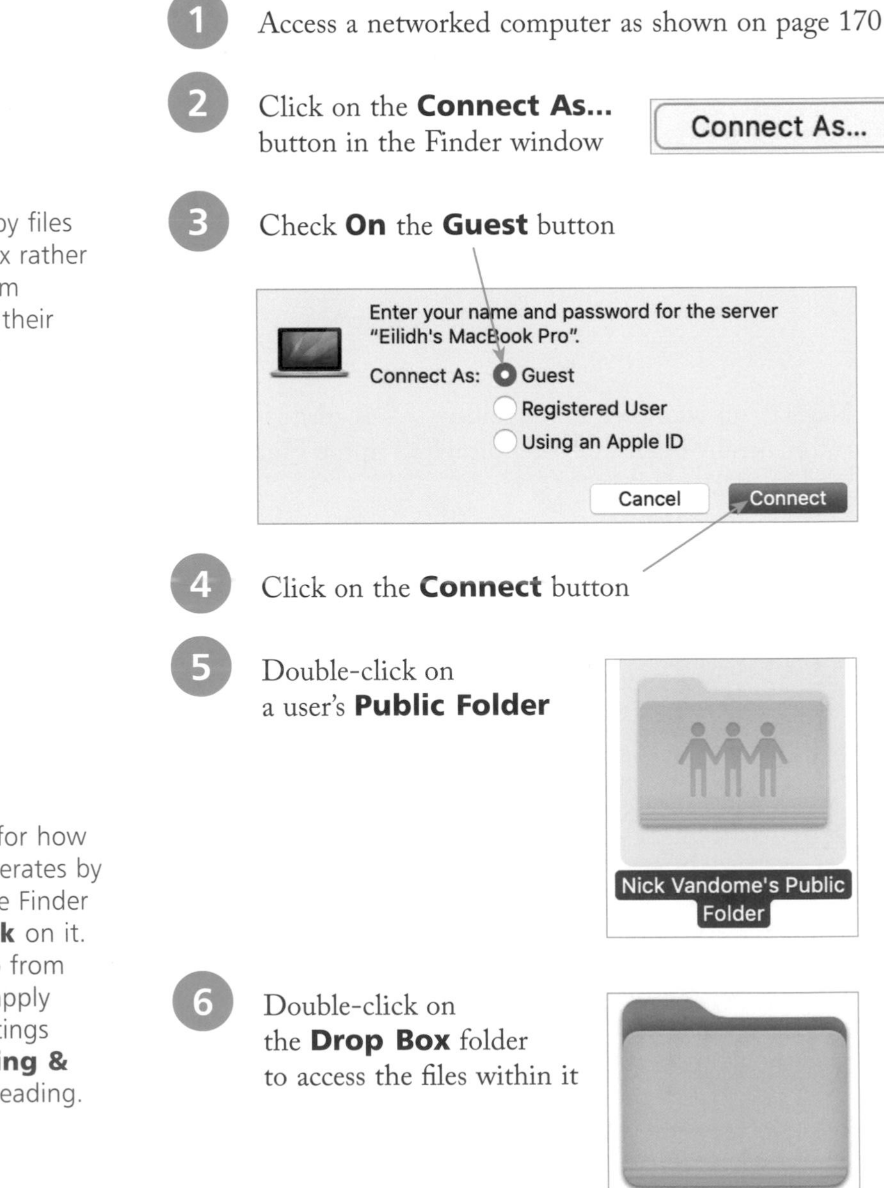

Set permissions for how the Drop Box operates by selecting it in the Finder and **Ctrl** + **click** on it. Select **Get Info** from the menu, and apply the required settings under the **Sharing & Permissions** heading.

11 Maintaining macOS

Despite its stability, macOS still benefits from a robust maintenance regime. This chapter looks at ways to keep macOS in top shape and ensure apps are as secure as possible, and looks at some general troubleshooting.

Time Machine

Time Machine is a feature of macOS that gives you great peace of mind. In conjunction with an external hard drive it creates a backup of your whole system, including folders, files, apps, and even the macOS operating system itself.

Once it has been set up, Time Machine takes a backup every hour, and you can then go into Time Machine to restore any files that have been deleted or become corrupt since the last backup.

Make sure that you have an external hard drive that is larger than the contents of your Mac, otherwise Time Machine will not be able to back it all up.

Setting up Time Machine

To use Time Machine, it first has to be set up. This involves attaching a hard drive to your Mac via cable or wirelessly, depending on the type of hard drive you have. To set up Time Machine:

1. Click on the **Time Machine** icon in **System Preferences**

2. In the Time Machine window, click on the **Select Disk...** button

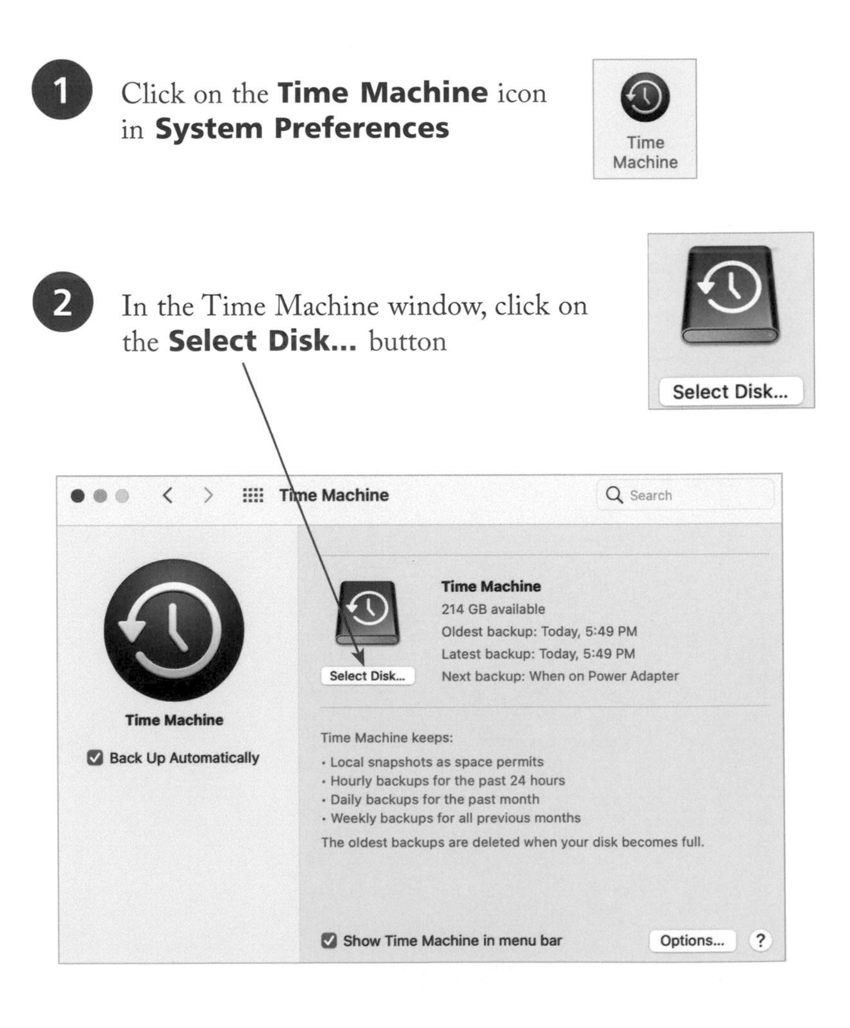

The hard drive must be connected in order to use Time Machine.

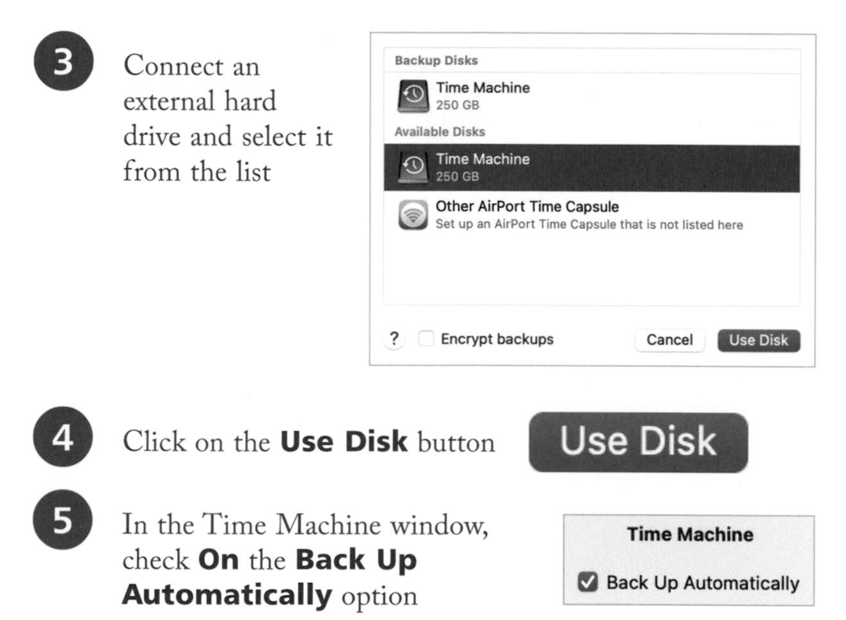

3. Connect an external hard drive and select it from the list

4. Click on the **Use Disk** button

Use Disk

5. In the Time Machine window, check **On** the **Back Up Automatically** option

Time Machine
Back Up Automatically

6. The backup will begin. The initial backup copies your whole system and can take several hours. Subsequent hourly backups only look at items that have been changed since the previous backup

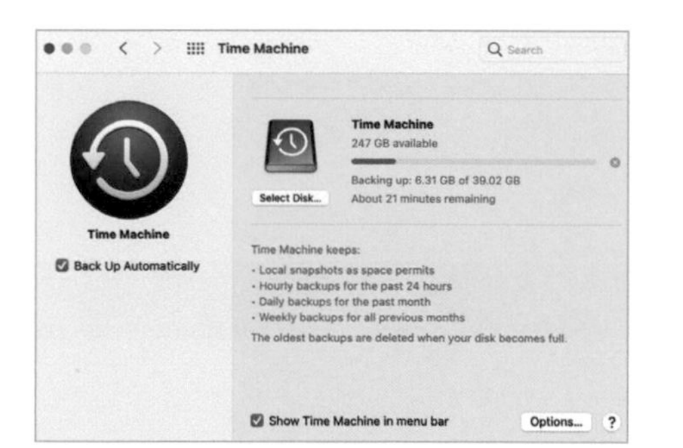

7. The record of backups and the schedule for the next one are shown here

When you first set up Time Machine it copies everything on your Mac. Depending on the type of connection you have for your external drive, this could take several hours. Because of this, it is a good idea to have a hard drive with a USB 3.0 or Thunderbolt connection to make it as fast as possible.

If you stop the initial backup before it has been completed, Time Machine will remember where it has stopped and resume the backup from this point.

...cont'd

Using Time Machine

Once Time Machine has been set up, it can then be used to go back in time to view items in an earlier state. To do this:

Access an item on your Mac and delete it. In this example, the folder **Time Machine 1** has been deleted

Click on the **Time Machine** icon on the Dock or in the Launchpad

Time Machine displays the selected window in its current state (the Time Machine 1 folder is deleted). Earlier versions are stacked behind it

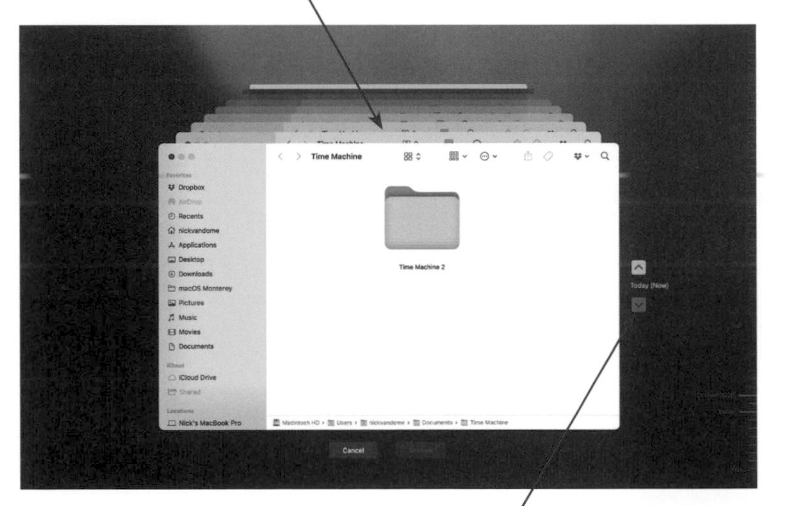

Click on the arrows to move through the open items, or select a time or date from the scale to the right of the arrows

Another way to move through Time Machine is to click on the pages behind the first one. This brings the selected item to the front

Beware

If you have deleted items before the initial setup of Time Machine, these will not be recoverable.

The active item that you were viewing before you launched Time Machine is the one that is active in the Time Machine interface. You can select items from within the active window to view their contents.

6. Move back through the windows to find the deleted item. Click on it and click on the **Restore** button to restore the item, in this case the **Time Machine 1** folder

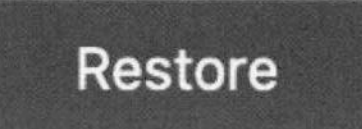

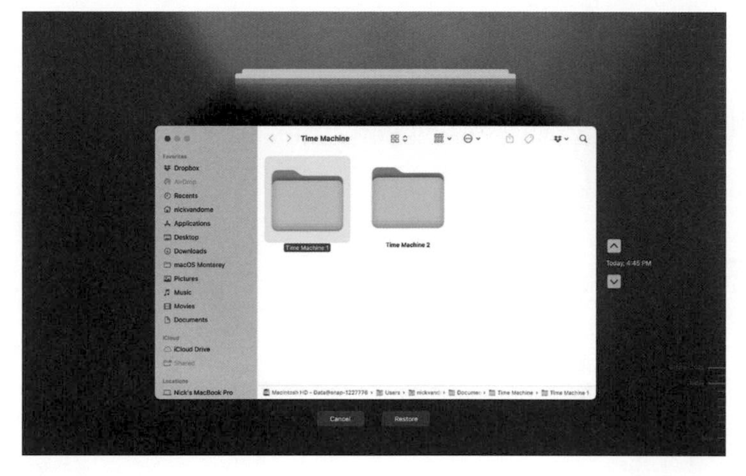

7. Click on the **Cancel** button to return to your normal environment, without restoring the item

Cancel

8. The deleted folder **Time Machine 1** is now restored to its original location

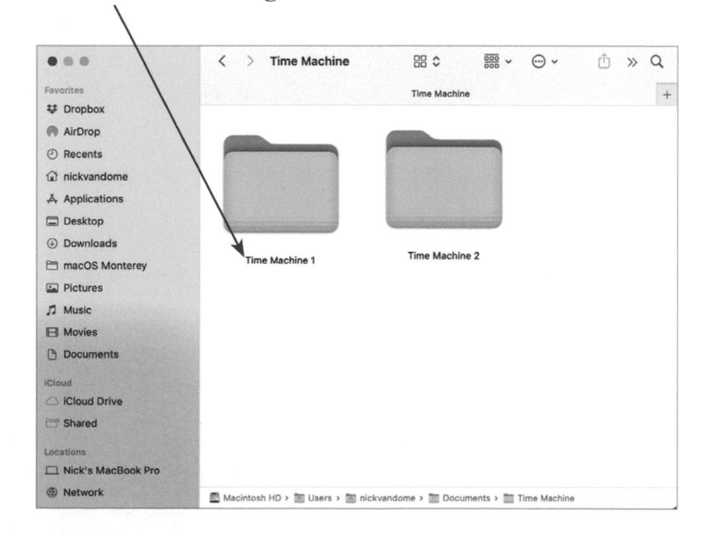

Items are restored from the Time Machine backup disk; i.e. the external hard drive.

Disk Utility

Disk Utility is a utility app that allows you to perform certain testing and repair functions for macOS. It incorporates a variety of functions, and is a good option for both general maintenance and if your computer is not running as it should.

Each of the functions within Disk Utility can be applied to specific drives and volumes. However, it is not possible to use the macOS start-up disk within Disk Utility as this will be in operation to run the app, and Disk Utility cannot operate on a disk that has apps already running. To use Disk Utility:

Disk Utility is located within the **Applications** > **Utilities (Other)** folder.

Checking disks

1. Open Disk Utility then click the **First Aid** tab to check a disk

2. Select a disk and select one of the First Aid options

If there is a problem with a disk and macOS can fix it, the **Repair** button will be available. Click on this to enable Disk Utility to repair the problem.

Erasing a disk

To erase all of the data on a disk or a volume:

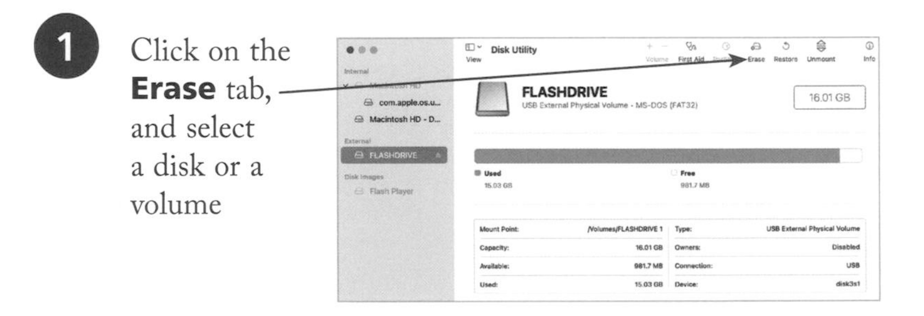

1. Click on the **Erase** tab, and select a disk or a volume

2. Click **Erase** to erase the data on the selected disk or volume

If you erase data from a removable disk such as a flashdrive, you will not be able to retrieve it.

System Information

This can be used to view how the different hardware and software elements on your Mac are performing. To do this:

Open the **Utilities** folder and double-click on the **System Information** icon

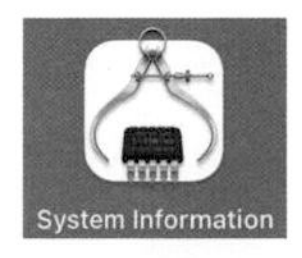

2 Click on the **Hardware** link and click on an item of hardware

Hardware
ATA
Apple Pay
Audio
Bluetooth
Camera
Card Reader
Controller
Diagnostics
Disc Burning
Ethernet Cards
Fibre Channel
FireWire
Graphics/Displays

3 Details about the item of hardware, and its performance, are displayed

Intel Iris Plus Graphics 640:

Chipset Model: Intel Iris Plus Graphics 640
Type: GPU
Bus: Built-In
VRAM (Dynamic, Max): 1536 MB
Vendor: Intel
Device ID: 0x5926
Revision ID: 0x0006
Metal: Supported, feature set macOS GPUFamily2 v1
Metal Family: Supported, Metal GPUFamily macOS 2
Displays:
Color LCD:
Display Type: Built-In Retina LCD
Resolution: 2560 x 1600 Retina
Framebuffer Depth: 24-Bit Color (ARGB8888)
Main Display: Yes
Mirror: Off
Online: Yes
Automatically Adjust Brightness: Yes
Connection Type: Internal

4 Similarly, click on network or software items to view their details

MacBook Pro

Fibre Channel
FireWire
Graphics/Displays
Memory
NVMExpress
PCI
Parallel SCSI
Power
Printers
SAS
SATA
Storage
Thunderbolt
USB
Network
Firewall
Locations
Volumes
WWAN
Wi-Fi

Firewall Settings:

Mode: Limit incoming connections to specific services and applications
Services:
File Sharing (SMB): Allow all connections
Printer Sharing: Allow all connections
Applications:
com.apple.WebKit.Networking: Allow all connections
com.getdropbox.dropbox: Allow all connections
Firewall Logging: Yes
Stealth Mode: No

System Information is located within the **Applications** > **Utilities (Other)** folder.

Activity Monitor

Activity Monitor is a utility app that can be used to view information about how much processing power and memory is being used to run apps. This can be useful to know if certain apps are running slowly or crashing frequently. To use Activity Monitor:

Activity Monitor is located within the **Applications** > **Utilities (Other)** folder.

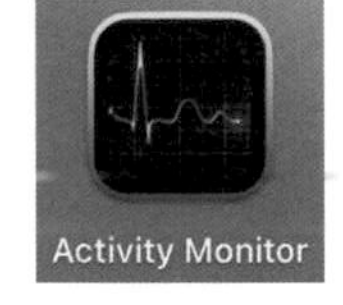

1. Open Activity Monitor and click on the **CPU** tab to see how much processor capacity is being used up

		CPU LOAD		
System:	3.12%		Threads:	1,990
User:	3.94%		Processes:	594
Idle:	92.94%			

2. Click on the **Memory** tab to see how much system memory (RAM) is being used up

MEMORY PRESSURE				
	Physical Memory:	8.00 GB	App Memory:	3.20 GB
	Memory Used:	6.41 GB	Wired Memory:	2.09 GB
	Cached Files:	1.48 GB	Compressed:	1.13 GB
	Swap Used:	2.34 GB		

3. Click on the **Energy** tab to see details of battery usage (for a laptop)

ENERGY IMPACT			BATTERY (Last 12 hours)
	Remaining charge:	31%	
	Time remaining:	1:58	
	Time on battery:	65:10	

4. Click on the **Disk** tab to see how much space has been taken up on the hard drive

		IO		
Reads in:	6,387,496		Data read:	193.32 GB
Writes out:	3,117,316		Data written:	121.86 GB
Reads in/sec:	0		Data read/sec:	819 bytes
Writes out/sec:	34		Data written/sec:	207 KB

Updating Software

Apple periodically releases updates for its software; both its apps and the macOS operating system. All of these are now available through the App Store. To update software:

1. Open **System Preferences** and click on the **Software Update** icon

2. Any available updates will be displayed. Click on the **Update Now** button to install an update

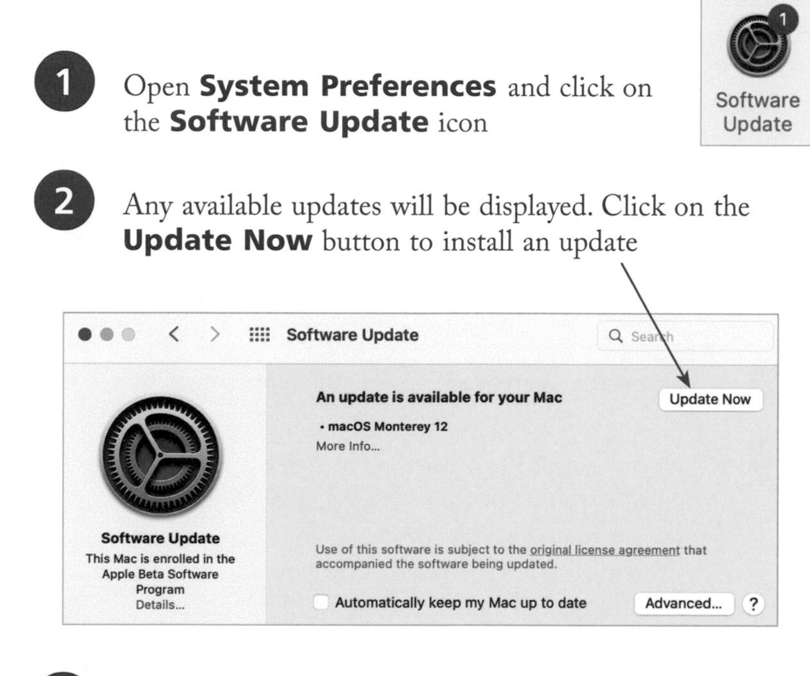

3. Click on the **Advanced...** button

Advanced...

4. Make the relevant selections for how updates are managed, including checking for them automatically and downloading new updates when they are available. Click on the **OK** button to confirm any changes

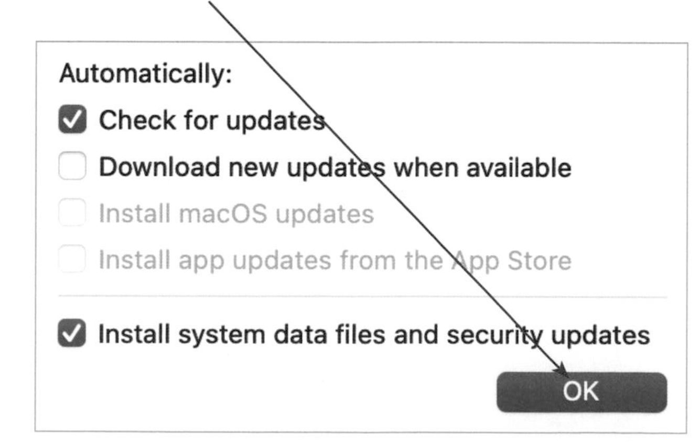

Hot tip

Software updates can also be accessed directly from the Apple menu, located on the top Menu bar. If updates are available this is denoted on the **App Store** link on the Apple menu, or on the App Store app's icon.

For some software updates, such as those for the macOS itself, you may have to restart your computer for them to take effect.

Gatekeeper

Internet security is an important issue for every computer user; no-one wants their computer to be infected with a virus or malicious software. Historically, Macs have been less prone to attack from viruses than Windows-based machines, but this does not mean Mac users can be complacent. With their increasing popularity there is now more temptation for virus writers to target them. macOS Monterey recognizes this, and has taken steps to prevent attacks with the Gatekeeper function. To use this:

Open **System Preferences** and click on the **Security & Privacy** icon

Click on the **General** tab

General

3. Click on one of these buttons to determine which location apps can be downloaded from. You can select from just the Mac App Store, or Mac App Store and identified developers, which gives you added security in terms of apps having been thoroughly checked

Allow apps downloaded from:
App Store
App Store and identified developers

Under the **General** tab there are also options for using a password when you log in to your account; if a password is required after sleep or if the screen saver is activated; showing a message when the screen is locked; or allowing your Apple Watch to unlock your Mac (check **Off** the **Disable automatic login** button)

General | FileVault | Firewall | Privacy
A login password has been set for this user Change Password...
Require password 1 hour after sleep or screen saver begins
Show a message when the screen is locked Set Lock Message...
Disable automatic login

To make changes within the **General** section of the Security & Privacy system preferences, click on the padlock icon and enter your admin password.

Click the lock to make changes.

If a password is not added for logging in to your account, or after sleep, other people could access your account and all of your files.

Privacy

Also within the Security & Privacy system preferences are options for activating a firewall and privacy settings. To access these:

1. Click on the **Firewall** tab
2. Click on the **Turn On Firewall** button to activate this. Click on **Firewall Options** to change settings for the firewall
3. Click on the **Privacy** tab
4. Click on the **Location Services** link and check **On** the **Enable Location Services** option if you want relevant apps to be able to access your location
5. Click on the **Contacts** link and check **On** any relevant apps that are allowed to access your contacts
6. Click on the **Analytics & Improvements** link and check **On** or **Off** the options for sharing analytical information with Apple and app developers. This will include any problems, and helps Apple improve its software and apps. This information is collected anonymously

Don't forget

A firewall is an application that aims to stop malicious software from accessing your computer.

Problems with Apps

The simple answer

macOS is something of a rarity in the world of computing software: it claims to be remarkably stable, and it is. However, this is not to say that things do not sometimes go wrong, although this is considerably less frequent than with older Mac operating systems. Sometimes this will be due to problems within particular apps, and on occasion the problems may lie with macOS itself. If this does happen, the first course of action is to restart macOS using the **Apple menu** > **Restart** command. If this does not work, or you cannot access the Restart command as the Mac has frozen, try turning off the power to the computer and then starting up again.

Force quitting

If a particular app is not responding, it can be closed down separately without the need to reboot the computer. To do this:

When there are updates to macOS, these can on rare occasions cause issues with some apps. However, these are usually fixed with subsequent patches and upgrades to macOS.

1. Select **Apple menu** > **Force Quit...** from the Menu bar

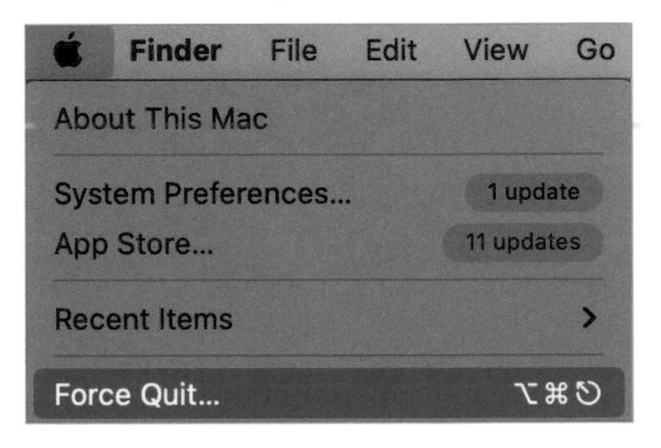

2. Select the app you want to close

3. Click the **Force Quit** button

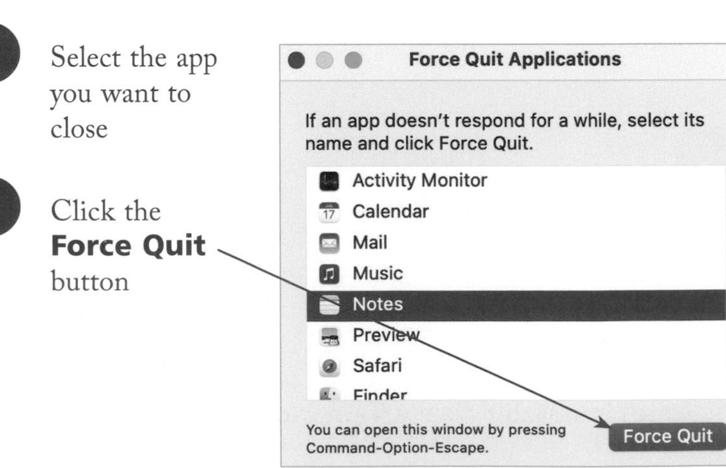

General Troubleshooting

It is true that things do occasionally go wrong with macOS, although probably with less regularity than with some other operating systems. If something does go wrong, there are a number of areas that you can check and also some steps you can take to ensure that you do not lose any important data if the worst-case scenario occurs, and your hard drive packs up completely:

- **Backup**. If everything does go wrong it is essential that you have taken preventative action in the form of making sure that all of your data is backed up and saved. This can be done with either the Time Machine app or by backing up manually by copying data to a flashdrive. Some content is also automatically backed up if you have iCloud activated.
- **Reboot**. One traditional reply by IT helpdesks is to reboot (i.e. turn off the computer and turn it back on again) and hope that the problem has resolved itself. In a lot of cases this simple operation does the trick, but it is not always a viable solution for major problems.
- **Check cables**. If the problem appears to be with a network connection or an externally-connected device, check that all cables are connected properly and have not become loose. If possible, make sure that all cables are tucked away so that they cannot be inadvertently pulled out.
- **Check network settings**. If your network or internet connections are not working, check the network settings in System Preferences. Sometimes when you make a change to one item, this can have an adverse effect on one of these settings. (If possible, lock the settings once you have applied them by clicking on the padlock icon in the Network preferences window.)
- **Check for viruses**. If your computer is infected with a virus this could affect the efficient running of the machine. Luckily, this is less of a problem for Macs as virus writers tend to concentrate their efforts toward Windows-based machines. However, this is changing as Macs become more popular, and there are plenty of Mac viruses out there. So, make sure your computer is protected by an app such as Norton AntiVirus, which is available from **www.norton.com**

In extreme cases, you will not be able to reboot your computer normally. If this happens, you will have to pull out the power cable and re-attach it. You will then be able to reboot, although the computer may want to check its hard drive to make sure that everything is in working order.

...cont'd

- **Check start-up items**. If you have set certain items to start automatically when your computer is turned on, this could cause certain conflicts within your machine. If this is the case, disable the items from launching during the booting-up of the computer. This can be done within the **Users & Groups** section of System Preferences by clicking on the **Login Items** tab, selecting the relevant item and pressing the minus button.

If you are having problems opening a document that you have been sent from a trusted source, contact them to make sure that the document has not been locked with a password. If you receive documents from someone you do not know, such as by email, do not open them as they may contain viruses or malware.

- **Check permissions**. If you or other users are having problems opening items, this could be because of the permissions that are set. To check these, select the item in the Finder, click on the **File** button on the top Menu bar and select **Get Info**. In the **Sharing & Permissions** section of the Info window you will be able to set the relevant permissions to allow other users, or yourself, to read, write or have no access.

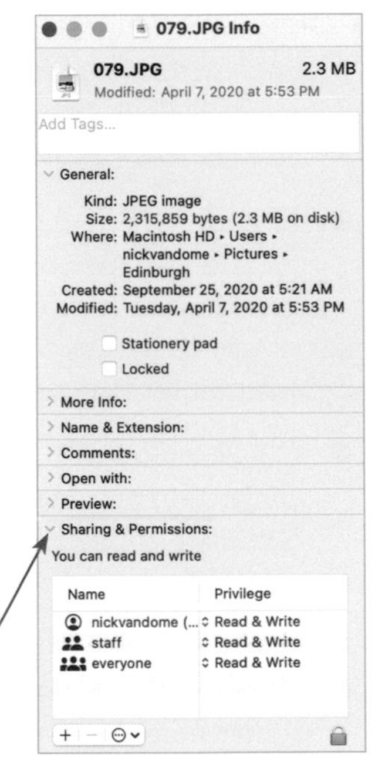

Click here to view permissions settings

Within the Screen Saver system preferences (Desktop & Screen Saver) is a button for **Hot Corners**. Click on this to access options for specifying what happens when the cursor is pointed over the four corners of the screen. This includes accessing the Notification Center and the Launchpad, or putting the Mac display to sleep.

- **Eject external devices**. Sometimes external devices, such as flashdrives, can become temperamental and refuse to eject the disks within them, or even not show up on the Desktop or in the Finder at all. If this happens, you can try to eject the disk by pressing the mouse button when the Mac chimes are heard during the booting-up process.
- **Turn off your screen saver**. Screen savers can sometimes cause conflicts within your computer, particularly if they have been downloaded from an unreliable source. If this happens, change the screen saver within the **Screen Saver** tab in the **Desktop & Screen Saver** preference of System Preferences, or disable it altogether.

Index

A

B

C

D

E

F

N

O

P

Q

R

S

T

U

V

W

Z